Advanced Mathematics
and
Mechanics Applications
Using
MATLAB

Howard B. Wilson
Louis H. Turcotte

CRC Press
Boca Raton   Ann Arbor   London   Tokyo

Library of Congress Cataloging-in-Publication Data

Catalog record is available from the Library of Congress

MATLAB is a trademark of The MathWorks, Inc.

No claim to original U.S. Government works
International Standard Book Number 0-8493-2482-3
Printed in the United States of America   3 4 5 6 7 8 9 0
Printed on acid-free paper

For my father who always loved learning.

Howard B. Wilson

For my loving wife, Evelyn, our cat, Patches, and my parents.

Louis H. Turcotte

Preface

This book uses MATLAB® to analyze various applications in mathematics and mechanics. The authors hope to encourage engineers and scientists to consider this modern programming environment as an excellent alternative to Fortran. MATLAB[1] embodies an interactive environment for technical computing which includes a high level programming language and remarkably simple graphics commands facilitating two- and three-dimensional data presentation. The wealth of intrinsic functions to handle matrix algebra, Fourier series, and complex valued functions makes straight forward calculator operations of many tasks previously requiring complicated subroutine libraries with lengthy and cumbersome argument lists.

The applications emphasize solutions of linear and nonlinear differential equations. Linear partial differential equations and linear matrix differential equations are analyzed using eigenfunctions and series solutions. Additionally, several other diverse topics, including conformal mapping, heat conduction, and inertial property calculation for areas are presented.

Structural dynamics of a cable system is used to illustrate the Runge-Kutta method for integrating systems of nonlinear differential equations. Because the writers have found the Runge-Kutta method adequate to handle most practical applications, other viable alternatives such as predictor-corrector methods, backward differences, or extrapolation are not treated. We have presented

---

[1]MATLAB is a registered trademark of The MathWorks, Inc. For additional information contact:

The MathWorks, Inc.
24 Prime Park Way
Natick, MA 01760-1500
(508) 653-1415, Fax: (508) 653-2997
Email: info@mathworks.com

one example of an Adams integrator involving a binary MEX-file (created using Fortran) of the well known public domain code ODE.

This book is not intended to be used as an introductory numerical analysis text. It is most useful as a reference book or a supplementary text in computationally oriented courses emphasizing physical applications. The authors have previously solved most of the examples in Fortran. Our MATLAB solutions consume about ninety pages (over six thousand lines). Although few books published recently present this much code, comparable Fortran versions would probably be four to five times as long. In fact, the conciseness achievable in MATLAB was a primary motivation for writing the book.

The programs have extensive comments and are intended for study as separate entities without an additional reference. Consequently, some deliberate redundancy exists between program comments and text discussions. We have also used a program listing style we feel will be helpful to most readers. The source listings have line numbers adjacent to the MATLAB code. MATLAB does not employ line numbers, nor does it incorporate use of infamous *goto* statements. Line numbers are included because they provide convenient reference points during discussions of particular program segments. We have also concatenated multiple MATLAB statements on the same line whenever possible without interrupting the logical flow.

All the programs presented are designed to operate under the 4.x version of MATLAB. The intended operating system environment is either Microsoft Windows or UNIX X-Windows, with both the text and graphics windows simultaneously being visible. A few of the programs not employing graphics may run without changes under MATLAB 3.5. However, utilization of the windowed operating environment and the many powerful graphics functions of MATLAB 4.x leads to programs requiring extensive changes to operate in MATLAB 3.5. Furthermore, all the programs are contained on a diskette accompanying the book. The disk is organized with directories corresponding to different chapters. A group of repeatedly used functions such as those for spline interpolation and interactive data input comprises a separate utility library. We believe that the programs contain a number of functions useful in other applications.

Finally, we emphasize that the primary audience we wish to reach involves people dealing with physical applications. A thorough grounding in ideas of Euclidean geometry, Newtonian mechanics, and some mathematics beyond calculus is essential to understand many of the topics. MATLAB is an exciting and extremely useful tool allowing increased efficiency for solving practical computing problems. We hope that this book encourages readers to entertain MATLAB as a highly attractive alternative to less robust software environments.

Howard B. Wilson

Louis H. Turcotte

Contents

List of Figures

List of Tables

List of MATLAB Examples

Advanced Mathematics
and
Mechanics Applications
Using
MATLAB

1

Introduction

1.1 MATLAB: A Tool for Engineering Analysis

This book illustrates various MATLAB applications in mechanics and applied mathematics. Our objective is to employ numerical methods in examples emphasizing the appeal of MATLAB as a programming tool. The programs are intended for study as a primary component of the text. The numerical methods include interpolation, numerical integration, finite differences, linear algebra, Fourier analysis, roots of nonlinear equations, linear differential equations, nonlinear differential equations, and linear partial differential equations. Many intrinsic functions included in MATLAB are used along with extensions developed by the authors. The physical applications vary widely from solution of linear and nonlinear differential equations in system dynamics to geometric property calculations for areas.

During the last twenty-five years FORTRAN has been the favorite language for solving mathematical and engineering problems on digital computers. FORTRAN is valuable for analyzing complex problems. Nevertheless, it is limited by an artificial and sometimes awkward mismatch between concise formulations typified by modern matrix methods and the FORTRAN code needed to implement such calculations. The new programming language MATLAB greatly reduces this computational barrier by providing a highly interactive medium, including many advanced mathematical tools as intrinsic functions. Advanced software features such as dynamic memory allocation and interactive error tracing reduce the time needed to develop problem solutions. The powerful but simple graphics commands in MATLAB also facilitate preparation of high quality graphs and surface plots appropriate for technical papers and books. Because the graphics features can be used interactively, most of the frustration encountered in debugging compiled FORTRAN code employing binary libraries is overcome. Experience of the authors indicates that a MATLAB problem solution which has some mathematical structure and employs graphics may require as little as one fifth as much code as that needed to accomplish the task in FORTRAN. Consequently, more time can be devoted to the primary purpose of computing, namely, improved understanding of physical system behavior.

Most of the mathematical background needed to understand the topics presented in this book is typically covered in an under-graduate engineering curriculum. This should include a thorough grounding in calculus, differential equations, and knowledge of a procedure oriented programming language such as FORTRAN. An additional course on advanced engineering mathematics covering linear algebra, matrix differential equations, and eigenfunction so-lutions of partial differential equations will also be valuable. The MATLAB programs in this book were written primarily to serve as instructional examples. The greatest benefit to the reader will probably be derived through careful study of the programs. Fur-thermore, we believe that several of the MATLAB functions are useful for practical applications. Typical examples include spline routines to interpolate, differentiate, and integrate; area and in-ertial moments for general plane shapes; and volume and inertial properties of arbitrary polyhedra.

The prospect for expanded use of advanced computing tools like MATLAB is bright. Continued growth will be fueled by a large community of users familiar with sophisticated analytical meth-ods. Furthermore, advances in software development techniques and remarkable decreases in hardware costs will accelerate these developments. The authors are hopeful that this book will moti-vate analysts already comfortable with traditional languages like FORTRAN to learn MATLAB. The rewards of such efforts will be considerable.

1.2   Use of MATLAB Commands and Related Reference Materials

MATLAB has a rich command vocabulary embracing numerous mathematical topics. The current section presents instructions on: a) how to learn MATLAB commands, b) how to examine and un-derstand MATLAB's lucidly written and easily accessible "demo" programs, and c) how to expand the command language by writ-ing new functions and programs. Table 1.2 (at the end of this chapter) contains a lengthy summary of the most useful operators and commands. Many other less familiar commands and functions are provided in auxiliary toolboxes. The reader is encouraged to study the command summary to get a feeling for the language structure and to have an awareness of powerful operations such as **null, orth, eig,** and **fft.**

The manual for *The Student Edition of MATLAB* [85] should be read thoroughly and kept handy for reference. Other references [40, 83, 86] also provide valuable supplementary information. This book expands the standard MATLAB documentation to include additional examples which we believe are complementary to more basic instructional materials.

Learning to use **help, type, demo,** and **diary** is an important first step to mastering MATLAB. **help** lists a summary of available commands with a short description of each command. **help** *function-name* (such as **help** *plot*) lists available documentation on a command or function generically called "function-name". MATLAB responds by printing internal comments contained at the start of the relevant function (comments are printed until the first blank or executable statement occurs). This feature allows users to create online help for functions they write by simply inserting appropriate comments at the top of the function. The instruction **type** *function-name* lists the entire source code for any function for which source code is available (the code for some intrinsic functions cannot be listed because it is stored in binary form for computational efficiency). Consider the following list of typical examples.

help help discusses use of help command.

help demos lists names of various demo programs.

type linspace lists the source code for the function which generates a vector of equidistant data values.

type plot causes an error condition and outputs a message indicating that **plot** is a built-in function.

intro executes the source code in a script file

called **intro** which illustrates various MAT-LAB functions.

type intro lists the source code for the **intro** demo program. By studying this example, readers can quickly learn to use many MAT-LAB commands.

graf2d demonstrates X-Y graphing.

graf3d demonstrates X-Y-Z graphing.

help diary provides instructions on how results appearing on the command screen can be saved into a file for later printing, editing, or merging with other text.

diary *fil_name* instructs MATLAB to record, into a file called *fil_name*, all text appearing on the command screen until the user types **diary off**. The **diary** command is especially useful for getting copies of library programs such as **zerodemo**. It can also be used to save a sequence of commands performed interactively. Then the saved text can be edited to create a MATLAB function for repeated use.

demo initiates access to a lengthy set of demo programs. Users should run all of the demos to fully appreciate the functionality of MATLAB. It is also helpful to obtain source listings of the demo programs. Programs used in the demos which should receive detailed study are: **intro**, **zerodemo**, **fitdemo**, **quaddemo**, **odedemo**, **ode45**, **fftdemo**, and **truss**. These functions and script files utilize most of the fundamental commands in MATLAB.

A program was written to illustrate various coding aspects in MATLAB. Users experienced with another procedure oriented language can learn to code in MATLAB by studying this instructional example along with other code segments included in the demo programs. Our example analyzes a problem of finding a root of a function when the function itself is defined by an integral. The analysis tools needed are a root finder and a numerical integrator. Although the intrinsic MATLAB root finder **fzero** and the integrator **quad8** are both excellent, we have included functions using interval halving and Simpson's rule to illustrate several programming features. The basic relationships occurring in the solution process are that the main program calls a root finder which repeatedly evaluates a function, the values of which are obtained by numerical integration. Our function is $F(x) = 1 - \int_1^x t^{-1} dt$. Since $F(x) = 1 - ln(x)$, it evidently equals zero when x equals e, the base of natural logarithms[1]. This value is compared with results obtained by the root finder. In addition to the numerical analysis process illustrated by this example, the program includes a handy function for reading several variables on a single line, and demonstrates the use of formatted output. The steps in the process are as follows:

1. Read search limits and error tolerances governing the root search.

2. Call a root finder (we use a simple bisection algorithm) to find a root of $F(x)$. The root finder must repeatedly evaluate $F(x)$ during the search.

3. To evaluate $F(x)$, a numerical integration routine (based on Simpson's rule) is called.

4. The numerical integration routine repeatedly evaluates the integrand $1/x$.

This process involves program modules nested five deep. The modules of the program are:

---

[1]MATLAB uses a value of e given by **exp**(1) = 2.71828182845904.

rootest	Read data, call root finder, print results.
bisect	Searches for a root between prescribed limits by interval halving.
fnc	Evaluate the function $1 - \int_1^x \frac{dt}{t}$.
simpson	Simpson rule integrator used to compute **fnc**.
oneovx	The integrand $1/x$ called by **simpson**.
read	A utility function which reads several data items on the same line.

A listing of the program appears below. We have utilized several different modes of data input and output for instructional purposes. The following notes direct attention to salient points in the program.

TABLE 1.1. Description of Code in Example

Routine	Line	Operation
rootest	1-21	Typing **help rootest** would print the first 21 comment lines
	25,26	Three global variables are declared and one initialized. Such variables should have unique names (appending an underscore to the variable name is a commonly employed method) to avoid unintentional conflict with local variables in other functions. Once a variable is declared global, it is known in other modules where these variables are also defined global.
	32-35	**fprintf** is used to write headings. Note use of ... to continue lines and \n to issue a line feed.
	37-43	Character strings are created for printing later.
	46	Start a loop which will continue until a **break** or **return** is reached. See line 54.
	51	Function **read** is used to input three variables.
	61-64	**fprintf** is used to print several lines requesting data input.
	67-70	**num2str** and **int2str** convert numbers to strings, which are used to compose a longer string containing data parameters.
		continued on next page

8

Routine	Line	Operation
continued from previous page		
	73-78	Logical sequence employing **if, else, end**.
	83	Use **cputime** to check computation time.
	85	Evaluate elapsed computation time.
	94-104	Print formatted results.
bisect	26	Specify global variables.
simpson	1	*funct* is a character string passed as the name of a function being integrated.
	22	**feval** is used to compute a function value.
	23,24	Vector indexing and **sum** are used to accumulate terms.
fnc	15	Use **round** to compute the number of integration subintervals.
oneovx	18	Increment a global variable to track the number of function evaluations.

1.2.1   EXAMPLE PROGRAM TO COMPUTE THE VALUE OF e

1.2.1.1   Output from Example

```
>> rootest


*** CALCULATING THE BASE OF NATURAL LOGARITHMS ***
***              BY ROOT SEARCH                 ***

Input lower search limit, upper search limit,
and an error tolerance for the root
(typical values are 2.5, 3.0, 1.e-5)
Use 0,0,0 to terminate execution
 ? > 2.5,3.0,1.e-5


Input the allowable number of iterations
(Use a negative value if intermediate root
estimates are to be printed)
? > -100


Search limits are from 2.5 to 3
Root tolerance = 1e-05 with max iterations = 100
```

i	e-approx	funct. val.
1	2.7500000000	-1.1601e-02
2	2.6250000000	3.4919e-02
3	2.6875000000	1.1389e-02
4	2.7187500000	-1.7222e-04
5	2.7031250000	5.5915e-03
6	2.7109375000	2.7055e-03
7	2.7148437500	1.2656e-03
8	2.7167968750	5.4643e-04
9	2.7177734375	1.8704e-04
10	2.7182617188	7.3980e-06
11	2.7185058594	-8.2413e-05
12	2.7183837891	-3.7509e-05
13	2.7183227539	-1.5056e-05
14	2.7182922363	-3.8288e-06
15	2.7182769775	1.7846e-06
16	2.7182846069	-1.0221e-06

```
Iterations             =  16
e-approximation        =  2.718280792
Percent error          =  -3.81205e-05
CPU time               =  0.738945 seconds
Integrand evaluations =  12977


Input lower search limit, upper search limit,
and an error tolerance for the root
(typical values are 2.5, 3.0, 1.e-5)
Use 0,0,0 to terminate execution
? > 2,4,1e-10


Input the allowable number of iterations
(Use a negative value if intermediate root
estimates are to be printed)
? > 100


Search limits are from 2 to 4
Root tolerance = 1e-10 with max iterations = 100


Iterations             =  34
e-approximation        =  2.718281829
Percent error          =  1.77576e-09
CPU time               =  0.554961 seconds
Integrand evaluations =  25403


Input lower search limit, upper search limit,
and an error tolerance for the root
(typical values are 2.5, 3.0, 1.e-5)
Use 0,0,0 to terminate execution
? > 0,0,0


All Done
```

1.2.1.2   Script File rootest

```
 1: % Example: rootest
 2: % ~~~~~~~~~~~~~~~~
 3: % This instructional program illustrates nested
 4: % function calls.  The base of natural
 5: % logarithms (e) is approximated by finding a
 6: % value of x which makes the function
 7: %
 8: %    fnc(x)=1-integral( (dt)/t ; t=1, => ,x )
 9: %
10: % equal zero.  The sequence of function calls
11: % is as follows: 1) The main program calls
12: % bisect to find roots by interval halving.
13: % 2) bisect calls fnc(x) which equals one
14: % minus the integral of 1/x over limits from
15: % 1 to x. 3) fnc(x) calls simpson to perform
16: % the numerical integration. 4) simpson
17: % calls the integrand oneovx = 1/x
18: %
19: % User m functions required:
20: %    bisect, fnc, simpson, oneovx, read
21: %-----------------------------------------------
22:
23: % Place selected parameters in global storage
24: % for accessibility by other routines
25: global kount_ ifprnt_ trouble_
26: trouble_=0;
27:
28: % Typical values for program parameters
29: % a=2.5; b=3.0; xtol=1.e-5; nmax/100;
30:
31: sgn_chng=0;
32: fprintf('\n** CALCULATING THE BASE OF ')
33: fprintf('NATURAL LOGARITHMS **')
34: fprintf('\n**                 BY ROOT SEARCH')
35: fprintf('                **\n\n')
36:
37: str1=['Input lower search limit, upper ', ...
38:       'search limit \nand an error '];
39: str2=['tolerance for the root \n(typical ', ...
40:       'values are 2.5, 3.0, 1.e-5)', ...
41:       '\nUse 0,0,0 to terminate execution'];
42: str3=['\nNo sign change between given ', ...
```

12

```
43:         'limits. Choose data again\n'];
44:
45: % Repeatedly read intervals for root search
46: while 1
47:    kount_=0;
48:    ifprnt_=0;
49:    while sgn_chng==0
50:      fprintf(str1); fprintf(str2);
51:      [a,b,xtol]=read('? > ');
52:      if norm([a,b,xtol]) == 0
53:        fprintf('\nAll Done\n');
54:        return;
55:      end
56:      sgn_chng=((fnc(a)*fnc(b)) <= 0 );
57:      if sgn_chng==0
58:        fprintf(str3);
59:      end
60:    end
61:    fprintf('\nInput the allowable number of ')
62:    fprintf('iterations \n(Use a negative value')
63:    fprintf(' if intermediate root \nestimates')
64:    fprintf(' are to be printed)');
65:    nmax=input('? > ');
66:    disp(' '),
67:    disp(['Search limits are from ', ...
68:        num2str(a),' to ',num2str(b)])
69:    s1=num2str(xtol); s2=int2str(abs(nmax));
70:    s3=int2str(abs(nmax));
71:    fprintf(['Root tolerance = ',s1])
72:    fprintf([' with max iterations = ',s2,'\n'])
73:    if nmax < 0
74:      ifprnt_=1;
75:      nmax=abs(nmax);
76:    else
77:      ifprnt_=0;
78:    end
79:    if ifprnt_==1
80:      fprintf('\n      i          e-approx')
81:      fprintf('       funct. val.')
82:    end
83:    cptim=cputime;
84:    [rt,ntotl]=bisect('fnc',a,b,xtol,nmax);
85:    cptim=cputime-cptim;
86:
87:    if trouble_==1
```

```
88:        fprintf(['\nRerun program using a ', ...
89:                  'larger number of iterations\n'])
90:     else
91: % Evaluate e to precision given by the computer
92:        acurat=exp(1);
93:        pcter=100*(rt-acurat)/acurat;
94:        fprintf('\n\nIterations           = ')
95:        fprintf('%g',ntotl)
96:        fprintf('\ne-approximation      = ')
97:        fprintf('%11.9f',rt)
98:        fprintf('\nPercent error        = ')
99:        fprintf('%g',pcter)
100:       fprintf('\nCPU time             = ')
101:       fprintf('%g seconds',cptim)
102:       fprintf('\nIntegrand evaluations = ')
103:       fprintf('%g',kount_)
104:       fprintf('\n\n')
105:    end
106:    sgn_chng=0;
107: end
```

1.2.1.3 Function bisect

```
1: function [rt,ntotl]=bisect(fn,a,b,xtol,nmax)
2: %
3: % [rt,ntotl]=bisect(fn,a,b,xtol,nmax)
4: % ~~~~~~~~~~~~~~~~~~~~~~~~~~~~~~~~~~~~
5: % Determine a root of function fn by interval
6: % halving
7: %
8: % fn      - character string giving the name of
9: %           the function for which a root is
10: %          sought
11: % a,b     - limits for the root search
12: % xtol    - uncertainty tolerance for the root
13: %           value
14: % nmax    - maximum allowable number of function
15: %           values to find the root to desired
16: %           accuracy
17: % rt      - computed approximation to the root
18: %           accurate within a tolerance of xtol
19: % ntotl   - number of function evaluations made
20: %           to compute the root
```

```
21: %
22: % User m functions called:   argument fn
23: %-------------------------------------------------
24:
25: % Global variables
26: global ifprnt_ trouble_
27:
28: % Check the number of function values needed to
29: % compute the root
30: ntotl=round(log(abs(a-b)/xtol)/log(2));
31: if ntotl > nmax
32:    str=['The root tolerance of ',  ...
33:        num2str(xtol),' takes '];
34:    str=[str,int2str(ntotl),' iterations.'];
35:    disp(str);
36:    str=['\nThis value exceeds the ' ...
37:        'allowed maximum of '];
38:    str=[str,int2str(nmax),' iterations\n'];
39:    fprintf(str);
40:    trouble_=1; return
41: end
42:
43: % Perform iteration to compute the root
44: fa=feval(fn,a);
45: for i=1:ntotl
46:    rt=(a+b)/2;
47:    frt=feval(fn,rt);
48:    if ifprnt_==1
49:      fprintf('\n  %4i    %13.10f    %13.4e', ...
50:              i, rt, frt)
51:    end
52:    if frt*fa > 0, a=rt; fa=frt; else, b=rt; end
53: end
54: rt=(a+b)/2;
```

1.2.1.4 Function fnc

```
1: function f=fnc(x)
2: %
3: % f=fnc(x)
4: % ~~~~~~~~
5: % This function integrates to get the function
6: % for which roots are to be computed.
```

```
7:  %
8:  % x     - argument vector for the function
9:  % f     - approximation for
10: %          f=1-integral( (dt)/t ; t=1, => ,x )
11: %
12: % User m functions called:  simpson, oneovx
13: %--------------------------------------------------
14:
15: ne=round(400*abs(x-1));
16: f=1-simpson('oneovx',1,x,ne);
```

1.2.1.5 Function oneovx

```
1:  function v=oneovx(x)
2:  %
3:  % v=oneovx(x)
4:  % ~~~~~~~~~~~~
5:  % This function is the integrand passed to
6:  % simpson. The number of function values is
7:  % passed as a global variable.
8:  %
9:  % x     - argument vector x
10: % v     - vector with components of 1./x
11: %
12: % User m functions called:  none.
13: %--------------------------------------------------
14:
15: global kount_
16:
17: v=1 ./x;
18: kount_=kount_+length(x);
```

1.2.1.6 Function read

```
1:  function [a1,a2,a3,a4,a5,a6,a7,a8,a9,a10, ...
2:            a11,a12,a13,a14,a15,a16,a17,a18, ...
3:            a19,a20]=read(labl)
4:  %
5:  % [a1,a2,a3,a4,a5,a6,a7,a8,a9,a10,a11,a12, ...
6:  %  a13,a14,a15,a16,a17,a18,a19,a20]=read(labl)
7:  % ~~~~~~~~~~~~~~~~~~~~~~~~~~~~~~~~~~~~~~~~~~~~~~~~~
```

```
 8: %
 9: % This function reads up to 20 variables on one
10: % line. The items should be separated by commas
11: % or blanks. Using more than 20 output
12: % variables will result in an error.
13: %
14: % labl                 - Label preceding the
15: %                        data entry.  It is set
16: %                        to '? ' if no value of
17: %                        labl is given.
18: % a1,a2,...,a_nargout  - The output variables
19: %                        which are created
20: %                        (cannot exceed 20)
21: %
22: % A typical function call is:
23: % [A,B,C,D]=read('Enter values of A,B,C,D: ')
24: %
25: % User m functions required: none
26: %-----------------------------------------------
27:
28: if nargin==0, labl='? '; end
29: n=nargout;
30: str=input(labl,'s'); str=['[',str,']'];
31: v=eval(str);
32: L=length(v);
33: if L>=n, v=v(1:n);
34:   else, v=[v,zeros(1,n-L)]; end
35: for j=1:nargout
36:   eval(['a',int2str(j),'=v(j);']);
37: end
```

1.3   Description of MATLAB Commands and Related Reference Materials

TABLE 1.2. Partial List of MATLAB 4.xx Statements

Managing Commands and Functions	
help	Online documentation.
what	Directory listing of M-, MAT- and MEX-files.
type	List M-file.
lookfor	Keyword search through the help entries.
which	Locate functions and files.
demo	Run demos.
path	Control MATLAB's search path.
info	Information about MATLAB and The MathWorks.
Managing Variables and the Workspace	
who	List current variables.
whos	List current variables, long form.
load	Retrieve variables from disk.
save	Save workspace variables to disk.
clear	Clear variables and functions from memory.
pack	Consolidate workspace memory.
size	Size of matrix.
length	Length of vector.
disp	Display matrix or text.
Working with Files and the Operating System	
cd	Change current working directory
dir	Directory listing.
delete	Delete file.
getenv	Get environment value.
!	Execute operating system command.
	continued on next page

diary	Save text of MATLAB session.
Controlling the Command Window	
clc	Clear command window.
home	Send cursor home.
format	Set output format.
echo	Echo commands inside script files.
more	Control paged output in command window.
Starting and Quitting from MATLAB	
quit	Terminate MATLAB.
ˆC	local abort
startup	M-file executed when MATLAB is invoked.
matlabrc	Master startup M-file.
Operators and Special Characters	
+	Plus.
−	Minus.
*	Matrix multiplication.
.*	Array multiplication.
∧	Matrix power.
.∧	Array power.
kron	Kronecker tensor product.
\	Backslash or left division.
/	Slash or right division.
./	Array division.
:	Subscripting, vector generation.
()	Parentheses.
[]	Brackets.
.	Decimal point.
..	Parent directory.
...	Continuation.
,	Comma.
;	Semicolon.
%	Comment.
	continued on next page

!	Exclamation point.
'	Transpose and quote.
.'	Nonconjugated transpose.
=	Assignment.
==	Equality.
<>	Relational operators.
&	Logical AND.
|	Logical OR.
~	Logical NOT.
xor	Logical EXCLUSIVE OR.
Logical Functions	
exist	Check if variables or functions exist.
any	True if any element or vector is true.
all	True if all elements of vector are true.
find	Find indices of non-zero elements.
isempty	True for empty matrices.
isstr	True for text string.
Matrix Operators	
+	Addition.
−	Subtraction.
*	Multiplication.
/	Right division.
\\	Left division.
∧	Power.
'	Conjugate transpose.
Array Operators	
+	Addition.
−	Subtraction.
.*	Multiplication.
./	Right division.
.\\	Left division.
.∧	Power.
	continued on next page

.'	Transpose.	
Relational and Logical Operators		
<	less than.	
<=	less than or equal.	
>	greater than.	
>=	greater than or equal.	
==	equal.	
~=	not equal.	
&	AND.	
|	OR.	
~	NOT.	
Special Characters		
[	Used to form vectors and matrices.	
]	See [.	
(	Arithmetic expression precedence.	
)	See (.	
,	Separate subscripts and function arguments.	
;	End rows, suppress printing.	
:	Subscripting, vector generation.	
!	Execute operating system command.	
MATLAB Programming Constructs		
function	Add new function.	
eval	Execute string with MATLAB expression.	
feval	Execute function specified by string.	
global	Define global variable.	
nargchk	Validate number of input arguments.	
Control Flow		
if	Conditionally execute statements.	
else	Used with **if**.	
elseif	Used with **if**.	
end	Terminate **if, for, while**.	
for	Repeat statements a specific number of times.	

continued on next page

while	Repeat statements an indefinite number of times.
break	Break out of **for** and **while** loops.
return	Return to invoking function.
error	Display message and abort function.
Interactive Input	
input	Prompt for user input.
keyboard	Invoke keyboard as if it were a script file.
menu	Generate menu of choices for user input.
pause	Wait for user response.
Elementary Matrices	
zeros	Zeros matrix.
ones	Ones matrix.
eye	Identity matrix.
rand	Uniformly distributed random numbers.
randn	Normally distributed random numbers.
linspace	Linearly spaced vector.
logspace	Logarithmically spaced vector.
meshgrid	X and Y arrays for 3-D plots.
:	Regularly spaced vector.
Special Variables and Constants	
ans	Most recent answer.
eps	Floating point relative accuracy.
realmax	Largest floating point number.
realmin	Smallest positive floating point number.
pi	π, 3.1415926535897...
i,j	$\sqrt{-1}$, imaginary unit.
inf	∞, infinity.
NaN	Not-a-Number.
flops	Count of floating point operations.
nargin	Number of function input arguments.
nargout	Number of function output arguments.

continued on next page

computer	Computer type.
Time and Dates	
clock	Wall clock.
cputime	Elapsed CPU time.
date	Calendar.
etime	Elapsed time function.
tic, toc	Stopwatch timer function.
Matrix Manipulation	
diag	Create or extract diagonals.
fliplr	Flip matrix in the left/right direction.
flipud	Flip matrix in the up/down direction.
reshape	Change size.
rot90	Rotate matrix 90 degrees.
tril	Extract lower triangular part.
triu	Extract upper triangular part.
:	Index into matrix, rearrange matrix.
Specialized Matrices	
compan	Companion matrix.
hadamard	Hadamard matrix.
hankel	Hankel matrix.
hilb	Hilbert matrix.
invhilb	Inverse Hilbert matrix.
magic	Magic square.
pascal	Pascal matrix.
rosser	Classical symmetric eigenvalue test problem.
toeplitz	Toeplitz matrix.
vander	Vandermonde matrix.
wilkinson	Wilkinson's eigenvalue test matrix.
Elementary Math Functions	
abs	Absolute value.
acos	Inverse cosine.
acosh	Inverse hyperbolic cosine.
	continued on next page

angle	Phase angle.
asin	Inverse sine.
asinh	Inverse hyperbolic sine.
atan	Inverse tangent.
atan2	Four quadrant inverse tangent.
atanh	Inverse hyperbolic tangent.
ceil	Round towards plus infinity.
conj	Complex conjugate.
cos	Cosine.
cosh	Hyperbolic cosine.
exp	Exponential.
fix	Round towards zero.
floor	Round towards minus infinity.
imag	Complex imaginary part.
log	Natural logarithm.
log10	Common logarithm.
real	Complex real part.
rem	Remainder after division.
round	Round towards nearest integer.
sign	Signum function.
sin	Sine.
sinh	Hyperbolic sine.
sqrt	Square root.
tan	Tangent.
tanh	Hyperbolic tangent.
Specialized Math Functions	
bessel	Bessel function.
ellipj	Jacobian elliptic integral.
ellipke	Complete elliptic integral.
erf	Error function.
gamma	Gamma function.
rat	Rational approximation.
continued on next page	

24

Matrix Analysis	
cond	Matrix condition number.
norm	Matrix or vector norm.
rcond	LINPACK reciprocal condition estimator.
rank	Number of linearly independent rows or columns.
det	Determinant.
trace	Sum of diagonal elements.
null	Null space.
orth	Orthogonalization.
rref	Reduced row echelon form.
Linear Equations	
**/ and **	Linear equation solution.
chol	Cholesky factorization.
lu	Factors from Gaussian elimination.
inv	Matrix inverse.
qr	Orthogonal-triangular decomposition.
nnls	Non-negative least-squares.
pinv	Pseudoinverse.
Eigenvalues and Singular Values	
eig	Eigenvalues and eigenvectors.
poly	Characteristic polynomial.
hess	Hessenberg form.
qz	Generalized eigenvalues.
rdf2csf	Real block diagonal form to complex diagonal form.
cdf2rdf	Complex-diagonal form to real block diagonal form.
schur	Schur decomposition.
balance	Diagonal scaling to improve eigenvalue accuracy.
svd	Singular value decomposition.

continued on next page

Matrix Functions	
expm	Matrix exponential.
logm	Matrix logarithm.
sqrtm	Matrix square root.
funm	Evaluate general matrix function.
Basic Operations	
max	Largest component.
min	Smallest component.
mean	Average or mean value.
median	Median value.
std	Standard deviation.
sort	Sort in ascending order.
sum	Sum of elements.
prod	Product of elements.
cumsum	Cumulative sum of elements.
cumprod	Cumulative product of elements.
trapz	Numerical integration using trapezoidal method.
Finite Differences	
diff	Difference function and approximate derivative.
gradient	Approximate gradient.
del2	Five-point discrete Laplacian.
Correlation	
corrcoef	Correlation coefficients.
cov	Covariance matrix.
Fourier Transforms	
fft	Discrete Fourier transform.
fft2	Two-dimensional discrete Fourier transform.
ifft	Inverse discrete Fourier transform.
ifft2	Two-dimensional inverse discrete Fourier transform.
	continued on next page

abs	Magnitude.
angle	Phase angle.
fftshift	Move zeroth lag to center of spectrum.
Polynomials	
roots	Find polynomial roots.
poly	Construct polynomial with specified roots.
polyval	Evaluate polynomial.
polyvalm	Evaluate polynomial with matrix argument.
residue	Partial-fraction expansion (residues).
polyfit	Fit polynomial to data.
conv	Multiply polynomials.
deconv	Divide polynomials.
Data Interpolation	
interp1	1-D interpolation (1-D table lookup).
interp2	2-D interpolation (2-D table lookup).
interpft	1-D interpolation using FFT method.
griddata	Data gridding.
Function Functions - Nonlinear Numerical Methods	
ode23	Solve differential equations, low order method.
ode45	Solve differential equations, high order method.
quad	Numerically evaluate integral, low order method.
quad8	Numerically evaluate integral, high order method.
fmin	Minimize function of one variable.
fmins	Minimize function of several variables.
fzero	Find zero of function of one variable.
fplot	Plot function.
Elementary X-Y Graphs	
plot	Linear plot.

continued on next page

loglog	Loglog scale plot.
semilogx	Semi-log scale plot.
semilogy	Semi-log scale plot.
fill	Draw filled 2-D polygons.
Specialized X-Y Graphs	
polar	Polar coordinate plot.
bar	Bar graph.
stairs	Stairstep plot.
errorbar	Error bar plot.
hist	Histogram plot.
fplot	Plot function.
Graph Annotation	
title	Graph title.
xlabel	X-axis label.
ylabel	Y-axis label.
zlabel	Z-axis label for 3-D plots.
text	Text annotation.
gtext	Mouse placement of text.
grid	Grid lines.
Line and Area Fill Commands	
plot3	Plot lines and points in 3-D space.
fill3	Draw filled 3-D polygons in 3-D space.
Contour and Other 2-D Plots of 3-D Data	
contour	Contour plot.
contour3	3-D contour plot.
Surface and Mesh Plots	
mesh	3-D mesh surface.
meshc	Combination mesh/contour plot.
surf	3-D shaded surface.
surfc	Combination surf/contour plot.
waterfall	Waterfall plot.
Graph Appearance	
continued on next page	

view	3-D graph viewpoint specification.
hidden	Mesh hidden line removal mode.
axis	Axis scaling and appearance.
3-D Objects	
cylinder	Generate cylinder.
sphere	Generate sphere.
Figure Window Creation and Control	
figure	Create figure (graph window).
gcf	Get handle to current figure.
clf	Clear current figure.
close	Close figure.
Axis Creation and Control	
subplot	Create axes in tiled positions.
axes	Create axes in arbitrary positions.
gca	Get handle to current axes.
cla	Clear current axes.
axis	Control axis scaling and appearance.
hold	Hold current graph.
Handle Graphics Objects	
figure	Create figure window.
axes	Create axes.
line	Create line.
text	Create text.
patch	Create patch.
surface	Create surface.
Handle Graphics Operations	
set	Set object properties.
get	Get object properties.
reset	Reset object properties.
delete	Delete object.
drawnow	Flush pending graphics events.
Hardcopy and Storage	

continued on next page

print	Print graph or save graph to file.
printopt	Configure local printer defaults.
orient	Set paper orientation.
General Character String Functions	
abs	Convert string to numeric values.
setstr	Convert numeric values to string.
isstr	True for string.
str2mat	Form text matrix from individual strings.
eval	Execute string with MATLAB expression.
String Comparison	
strcmp	Compare strings.
upper	Convert string to uppercase.
lower	Convert string to lowercase.
String to Number Conversion	
num2str	Convert number to string.
int2str	Convert integer to string.
str2num	Convert string to number.
sprintf	Convert number to string under format control.
sscanf	Convert string to number under format control.

2

Elementary Aspects of MATLAB Graphics

2.1 Introduction

MATLAB's capabilities for plotting curves and surfaces are versatile and easy to understand. In fact, the effort required to learn MATLAB would prove worthwhile if students only needed to construct plots, save graphic images, and output publication quality graphs on a laser printer. Numerous help features and well-written demo programs are included with MATLAB. By executing the demo programs and studying the relevant code, users can quickly understand the techniques necessary to implement graphics within their programs. This chapter discusses only a few of the graphics commands. These commands are useful in many applications and do not require extensive time to master. This next section will provide a quick overview of the basics of using MATLAB's graphics. The subsequent sections in this chapter present four additional examples (summarized in the table below) involving interesting applications which use these graphics primitives.

Example	Purpose
Polynomial Interpolation	2-D graphics and polynomial interpolation functions
Conformal Mapping	2-D graphics and some aspects of complex numbers
String Vibration	2-D and 3-D graphics for a function of form $y(x, t)$
Animation of a Rotating Cube	3-D graphics showing animated rotation of a cube about an axis

2.2   Overview of Graphics

The following commands should be executed since they will accelerate the understanding of graphics functions, and others, included within MATLAB.

help help	discusses use of **help** command.
help	lists categories of help.
help general	lists various utility commands.
help more	describes how to control output paging.
help diary	describes how to save console output to a file.
help plotxy	describes 2D plot functions.
help plotxyz	describes 3D plot functions.
help graphics	describes more general graphics features.
help demos	lists names of various demo programs.
intro	executes the **intro** program showing MATLAB commands including fundamental graphics capabilities.
help funfun	describes several numerical analysis programs contained in MATLAB.
type humps	lists a function employed in several of the MATLAB demos.
fplotdemo	executes program **fplotdemo** which plots the function named **humps**.
help peaks	describes a function **peaks** used to illustrate surface plots.
peaks	executes the function **peaks** to produce an interesting surface plot.
splinc2d	executes a demo program to draw a curve through data input interactively.

The example programs can be studied interactively by issuing the command **more on** and then using the **type** command to list programs of interest. Furthermore, a code listing can be saved in a diary file for routing to any convenient print device. For example, making a source listing of the demo program **spline2d** is achieved by the following command sequence:

```
diary splin2d.sav;
type splin2d;
diary off
```

This creates, in the current working directory, an ASCII file named *splin2d.sav*. The user may then manipulate this file appropriately (e.g. edit, print, etc.).

More advanced features of MATLAB graphics, including handle graphics, control of shading and light sources, creation of movies, etc., exceed the scope of the present text. Instead we concentrate on using the basic commands listed below and on producing simple animations. The more advanced graphics features can be mastered by studying the MATLAB manuals and relevant demo programs.

The principal graphing commands discussed here are

Command	Purpose
plot	draw two-dimensional graphs
xlabel, ylabel zlabel	define axis labels
title	define graph title
axis	set various axis parameters (min, max, etc.)
text	place text at selected locations
grid	draw grid lines
mesh	draw surface plot with mesh
surface	draw surface plot
hold	fix the graph limits between successive plots
view	change surface viewing position
drawnow	empty graphics buffer immediately
cla	clear graphics window
contour	draw contour plot
ginput	read coordinates interactively

All of these commands, along with numerous others are extensively documented by the help facilities in MATLAB. The user can get an introduction to these capabilities by typing "**help plot**" and by running the demo programs. The accompanying code for the demo program should be examined since it provides worthwhile insight into how MATLAB graphics is used.

2.3 Polynomial Interpolation Example

Many familiar mathematical functions such as $\arctan(x)$, $\exp(x)$, $\sin(x)$, etc. can be represented well near $x = 0$ by Taylor series expansions. If a series expansion converges rapidly, taking a few terms in the series may yield good polynomial approximations. Assuming such a procedure is plausible, one approach to polynomial approximation results by taking some data points, say (x_i, y_i), $1 \leq i \leq n$ and determining the polynomial of degree $(n - 1)$ which passes through those points. It appears reasonable that using evenly spaced data is appropriate and that increasing the number of polynomial terms should improve the accuracy of the approximating function. However, it has actually been found that passing a polynomial through points on a function $y(x)$, where x values are evenly spaced, typically gives approximations which are not smooth between the data points and tend to oscillate at the ends of the interpolating interval [17]. Attempting to reduce the oscillation by raising the polynomial order makes matters worse. Therefore, a special set of unevenly spaced points bunching data near the interval ends according to

$$x_j = \frac{a + b}{2} + \left(\frac{a - b}{2}\right) \cos\left[\frac{\pi}{n}\left(j - \frac{1}{2}\right)\right] \qquad 1 \leq j \leq n$$

for the interval $a \leq x \leq b$ is preferable. This formula gives the so-called Chebyshev points which are optimally chosen in the sense described by Conte and de Boor [17].

The interpolation program **polyplot** employs MATLAB functions **polyfit** and **polyval** to compute approximations of the function $y = 1/(1 + x^4)$. Function **polyfit** evaluates coefficients in a polynomial of degree n which fits m data points by least square approximation. For a unique solution we need $m \geq n + 1$. The polynomial fits the data perfectly when $m = n + 1$. The function **polyval** can then be used to evaluate the polynomials for arbitrary arguments. The data points are computed using function **linspace** for equidistant data and **chbpts** for Chebyshev points. The graphic functions used in this example are **plot**, **title**, **xlabel**, **ylabel**, **get**, and **text**. It is worthwhile to study selected lines in the program. These lines and their operation are summarized below.

Line	Operation
20	**xe = xe**(:) insures **xe** is a column
24	**ye = 1./(1 + xe.** $\wedge$ **p)** raises each component of **xe** to power **p**, adds one and reciprocates the results componentwise.
50-52	creates a string which is continued between lines by using ...
53	**title** prints string **titl**. Shorter strings can be directly included as an argument in the function.
60	function **get** is used to find the range of the existing plot scale
66	**text** places a label at an appropriate interior point of the graph.
73	**genprint** saves the graphics image to a file for later printing.

The program produces the graph in Figure 2.1. Notice that the interpolation function through equidistant data points oscillates severely near ± 4, whereas the Chebyshev points produce more reasonable results. By changing the program data the reader can verify that using more interpolation points improves the Chebyshev approximation but makes the results from equidistant data much worse.

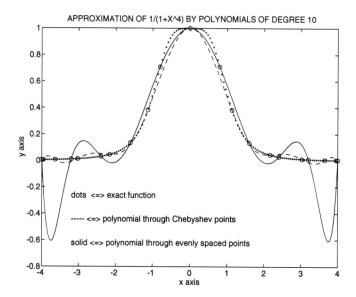

FIGURE 2.1. Approximation by Polynomials of Degree 10

MATLAB EXAMPLE

2.3.0.1 Script File polyplot

```
 1: % Example:  polyplot
 2: % ~~~~~~~~~~~~~~~~~~~~
 3: % This program illustrates the use of various
 4: % graphics commands to show how the location
 5: % of interpolation points affects the accuracy
 6: % and smoothness of polynomial approximations
 7: % to the function 1/(1+x^p).
 8: %
 9: % User m functions required:
10: %    chbpts
11: %------------------------------------------------
12:
13: % Set data parameters. Functions linspace and
14: % chbpts generate data with even and Chebyshev
15: % spacing
16: clear; n=11; a=-4; b=4; p=4;
17: xe=linspace(a,b,n); xc=chbpts(a,b,n);
18:
19: % Make sure all vectors are columns
20: xe=xe(:); xc=xc(:);
21: xx=linspace(a,b,151); xx=xx(:);
22:
23: % Compute function values
24: ye=1 ./(1+xe.^p);
25: yc=1 ./(1+xc.^p);
26: yy=1 ./(1+xx.^p);
27:
28: % Use function polyfit to compute polynomial
29: % coefficients.  Since the number of data
30: % points is one greater than the polynomial
31: % order, the polynomials will pass exactly
32: % through the data points.
33: cofe=polyfit(xe,ye,n-1);
34: cofc=polyfit(xc,yc,n-1);
35:
36: % Use function polyval to evaluate the
37: % polynomials
38: yye=polyval(cofe,xx); yyc=polyval(cofc,xx);
39:
40: % Plot the exact function and interpolating
```

```
41: % polynomials. Note that ... is used to
42: % continue a line
43: plot(xx,yy,'w.',xx,yye,'w-',xx,yyc,'w--',...
44: xe,ye,'wo',xc,yc,'wo')
45:
46: % Form a title and place it on the graph. Note
47: % that the functions num2str and int2str
48: % convert real and integer numbers into
49: % strings.
50: titl=['APPROXIMATION OF 1/(1+X^', ...
51:     num2str(p),') BY ',...
52:     'POLYNOMIALS OF DEGREE ',int2str(n-1)];
53: title(titl)
54:
55: % Place labels on the x and y axes
56: xlabel('x axis'); ylabel('y axis');
57:
58: % Use graphics functions gca and get to compute
59: % positions where text labels are to be placed
60: h=gca; vx=get(h,'XLim'); vy=get(h,'YLim');
61: xmin=vx(1); xmax=vx(2); dx=xmax-xmin;
62: ymin=vy(1); ymax=vy(2);
63: dy=ymax-ymin; xl=xmin+.1*dx;
64:
65: % Add labels to complete the graph
66: text(xl,ymin+.3*dy,'dots  <=> exact function')
67: text(xl,ymin+.2*dy,...
68:     ['----- <=> polynomial through ' ...
69:      'Chebyshev points'])
70: text(xl,ymin+.1*dy,...
71:     ['solid <=> polynomial through ' ...
72:      'evenly spaced points'])
73: genprint('polyplot')
74: disp(' '), disp('All Done')
```

2.3.0.2 Function chbpts

```
1: function x=chbpts(xmin,xmax,n)
2: %
3: % x=chbpts(xmin,xmax,n)
4: % ~~~~~~~~~~~~~~~~~~~~~
5: % Determine n points with Chebyshev spacing
6: % between xmin and xmax.
```

```
 7: %
 8: % User m functions called:  none
 9: %------------------------------------------------
10:
11: x=(xmin+xmax)/2+((xmin-xmax)/2)* ...
12:    cos(pi/n*((0:n-1)'+.5));
```

2.3.0.3   Function genprint

```
 1: function genprint(fname,append)
 2: %
 3: % genprint(fname,append)
 4: % ~~~~~~~~~~~~~~~~~~~~~~~
 5: % This function saves a plot to a file.
 6: %
 7: % fname  - name of file to save plot to
 8: % append - optional, if included plot is
 9: %          appended to file fname
10: %
11: % SYSTEM DEPENDENT ROUTINE
12: %
13: % User m functions called:  removfil
14: %------------------------------------------------
15:
16: % Save to postscript file
17: if nargin == 1
18:    eval(['removfil ', fname, '.ps']);
19:    eval(['print -dps ',fname]);
20: else
21:    eval(['print -dps -append ',fname]);
22: end
```

2.4   Conformal Mapping Example

This example involves analytic functions and conformal mapping. The complex function $w(z)$ which maps $|z| \leq 1$ onto the interior of a square of side length 2 can be written in power series form as

$$w(z) = \sum_{k=0}^{\infty} b_k z^{4k+1}$$

where

$$b_k = c\left[\frac{(-1)^k(\frac{1}{2})_k}{k!(4k+1)}\right] \qquad \sum_{k=0}^{\infty} b_k = 1$$

and c is a scaling coefficient chosen to make $z = 1$ map to $w = 1$ [64]. Truncating the series after some finite number of terms, say m, produces an approximate square with rounded corners. Increasing m reduces the corner rounding but convergence is rather slow so that using even a thousand terms still gives perceptible inaccuracy. The purpose of the present exercise is to show how a polar coordinate region characterized by

$$z = re^{i\theta} \qquad r_1 \leq r \leq r_2 \qquad \theta_1 \leq \theta \leq \theta_2$$

transforms and to exhibit an undistorted plot of the region produced in the w-plane. The exercise also emphasizes the utility of MATLAB for handling complex arithmetic and complex functions. The program has a short driver **squarrun** and a function **squarmap** which computes points in the w region and coefficients in the series expansion. Salient features of the program are summarized below.

Routine	Line	Operation
squarrun	14-40	functions **input, disp, fprintf,** and **read** are used to input data interactively. Several different methods of printing were used for purposes of illustration rather than necessity.
	43	function **squarmap** generates results.
	47	function **genprint** is a system dependent routine which is used to create plot files for later printing.
squarmap	31-33	functions **linspace** and **ones** are used to generate points in the z-plane.
	43-45	series coefficients are computed using **cumprod** and the mapping is evaluated using **polyval** with a matrix argument.
	48-51	scale limits are calculated to allow an undistorted plot of the geometry. Use is made of MATLAB functions **real** and **imag**.
	59-70	loops are executed to plot the circumferential lines first and the radial lines second.

Results produced when $0.5 \leq r \leq 1$ and $0 \leq \theta \leq 2\pi$ by a twenty-term series appear in Figure 2.2. The reader may find it interesting to run the program using several hundred terms and take $0 \leq \theta \leq \pi/2$. The corner rounding remains noticeable even when $m = 1000$ is used. Later in this book we will visit the mapping problem again to show that a better approximation is obtainable using rational functions.

42

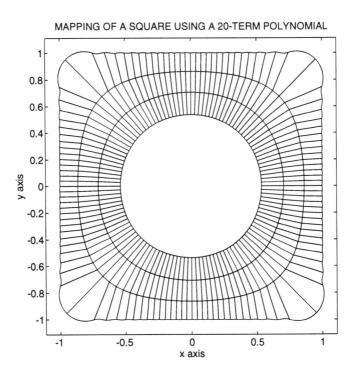

FIGURE 2.2. Mapping of a Square Using a 20-Term Polynomial

MATLAB EXAMPLE

2.4.0.4   Script File squarrun

```
 1: % Example:  squarrun
 2: % ~~~~~~~~~~~~~~~~~~~~
 3: % Driver program to plot the mapping of a
 4: % circular disk onto the interior of a square
 5: % by the Schwarz-Christoffel transformation.
 6: %
 7: % User m functions required:
 8: %    squarmap, read, removfil, genprint
 9: %-------------------------------------------------
10:
11: % Illustrate use of the function input to input
12: % data on the same line
13:
14: fprintf('\nCONFORMAL MAPPING OF A SQUARE ')
15: fprintf('BY USE OF A\n')
16: fprintf('TRUNCATED SCHWARZ-CHRISTOFFEL ')
17: fprintf('SERIES\n\n')
18:
19: fprintf('Input the number of series ')
20: fprintf('terms used')
21: m=input('(try 20)? ');
22:
23: % Illustrate use of the function disp
24: disp('')
25: str=['\nInput the inner radius, outer ' ...
26:      'radius and number of increments ' ...
27:      '\n(try .5,1,4)'];
28: fprintf(str);
29:
30: % Use function read to input several variables
31: [r1,r2,nr]=read;
32:
33: % Use function fprintf to print more
34: % complicated heading
35: str=['\nInput the starting value of ' ...
36:      'theta, the final value of theta \n' ...
37:      'and the number of theta increments ' ...
38:      '(the angles are in degrees) ' ...
39:      '\n(try 0,360,120)'];
40: fprintf(str); [t1,t2,nt]=read;
```

```
41:
42: % Call function squarmap to make the plot
43: [w,b]=squarmap(m,r1,r2,nr,t1,t2,nt+1);
44:
45: % Remove an existing plot file if it exists.
46: % Then save the plot.
47: genprint('squarplt');
48: disp(' '); disp('All Done')
```

2.4.0.5 Function squarmap

```
 1: function [w,b]=squarmap(m,r1,r2,t1,t2,nt)
 2: %
 3: % [w,b]=squarmap(m,r1,r2,nr,t1,t2,nt)
 4: % ~~~~~~~~~~~~~~~~~~~~~~~~~~~~~~~~~~~~~~~~
 5: % This function evaluates the conformal mapping
 6: % produced by the Schwarz-Christoffel
 7: % transformation w(z) mapping abs(z)<=1 inside
 8: % a square having a side length of two.  The
 9: % transfomation is approximated in series form
10: % which converges very slowly near the corners.
11: %
12: % m         - number of series terms used
13: % r1,r2,nr - abs(z) varies from r1 to r2 in
14: %             nr steps
15: % t1,t2,nt - arg(z) varies from t1 to t2 in
16: %             nt steps (t1 and t2 are measured
17: %             in degrees)
18: % w         - points approximating the square
19: % b         - coefficients in the truncated
20: %             series expansion which has the
21: %             form
22: %
23: %          w(z)=sum({j=1:m},b(j)*z*(4*j-3))
24: %
25: % User m functions called:  none
26: %-------------------------------------------------
27:
28: % Generate polar coordinate grid points for the
29: % map.  Function linspace generates vectors
30: % with equally spaced components.
31: r=linspace(r1,r2,nr)';
32: t=pi/180*linspace(t1,t2,nt);
```

```
33: z=(r*ones(1,nt)).*(ones(nr,1)*exp(i*t));
34:
35: % Use high point resolution for the
36: % outer contour
37: touter=pi/180*linspace(t1,t2,10*nt);
38: zouter=r2*exp(i*touter);
39:
40: % Compute the series coefficients and
41: % evaluate the series
42: k=1:m-1;
43: b=cumprod([1,-(k-.75).*(k-.5)./(k.*(k+.25))]);
44: b=b/sum(b); w=z.*polyval(b(m:-1:1),z.^4);
45: wouter=zouter.*polyval(b(m:-1:1),zouter.^4);
46:
47: % Set graph limits for an undistorted plot
48: uu=real([w(:);wouter(:)]);
49: vv=imag([w(:);wouter(:)]);
50: umin=min(uu); umax=max(uu); uc=(umin+umax)/2;
51: vmin=min(vv); vmax=max(vv); vc=(vmin+vmax)/2;
52: d=max(umax-umin,vmax-vmin)*.55; hold off; cla;
53: axis('square');
54: axis([uc-d;uc+d;vc-d;vc+d]); hold on
55:
56: % Plot orthogonal grid lines which represent
57: % the mapping of circles and radial lines
58:
59: % First draw the circle maps
60: for j=1:nr-1
61:    wj=w(j,:);
62:    plot(real(wj),imag(wj),'w');
63: end
64: plot(real(wouter),imag(wouter),'w')
65:
66: % Then draw the radial line maps
67: for j=1:nt
68:    wj=w(:,j);
69:    plot(real(wj),imag(wj),'w');
70: end
71:
72: % Add a title and axis labels
73: title(['MAPPING OF A SQUARE USING A ', ...
74:         num2str(m),'-TERM POLYNOMIAL'])
75: xlabel('x axis'); ylabel('y axis')
76: hold off
```

2.5 String Vibration Example

The next example investigates a familiar type of problem involving a function $y(x, t)$ known for $a \leq x \leq b$ and $t > 0$. For the case studied, y represents transverse deflection in a taut string. However, it could just as well describe current flow in a transmission line or pressure in a porous medium.

In the string vibration problem, we use dimensionless variables to simplify interpreting the results. The string is at rest initially. The left end is fixed and the right end is suddenly given a sinusoidally varying motion causing waves which propagate back and forth along the string. The related boundary value problem in dimensionless variables is

$$\frac{\partial^2 y}{\partial x^2} = \frac{\partial^2 y}{\partial t^2} \qquad 0 \leq x \leq 1 \qquad t > 0$$

$$y(0, t) = 0 \qquad y(1, t) = \sin(\omega t)$$

$$y(x, 0) = 0 \qquad \frac{\partial y}{\partial t}(x, 0) = 0$$

As long as ω does not equal an integer multiple of π, this problem has a simple series solution given by

$$y(x, t) \quad = \quad \frac{\sin(\omega x)\sin(\omega t)}{\sin(\omega)} - \frac{\omega}{\pi \sin(\omega)} \{$$

$$\sum_{n=1}^{\infty} \left[\frac{\sin(\omega - n\pi)}{n(\omega - n\pi)} - \frac{\sin(\omega + n\pi)}{n(\omega + n\pi)} \right] \sin(n\pi t)\sin(n\pi x) \}$$

We need graphical results to help us visualize the dynamics of the system. Note that at any fixed position x_0, $y(x_0, t)$ represents the time history of that point on the string. Similarly, at any particular time t_0, then $y(x, t_0)$ for $0 \leq x \leq 1$ represents the deflection pattern of the string. Suppose $\omega \approx k\pi$ for integer k, then a nearly zero denominator occurs in the solution, thereby producing large displacement amplitudes. This results in transverse deflections which build up from zero initially to large values as translating waves repeatedly reflect from the end boundaries of the string. Terms in the solution such as

$$\sin(n\pi x)\sin(n\pi t) = \frac{1}{2}\{\cos[n\pi(x - t)] - \cos[n\pi(x + t)]\}$$

MOTION OF A STRING HAVING ONE END SHAKEN HARMONICALLY

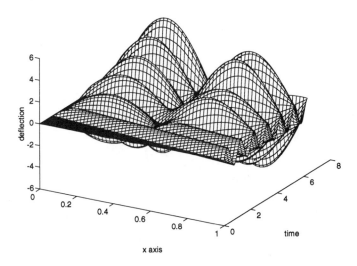

FIGURE 2.3. String Motion: One End is Shaken Harmonically

represent components which simultaneously translate to the left and to the right with unit speed. Consequently, no displacement will occur near the end $x = 0$ until $t \geq 1$. By the time $t = 10$, for example, several reflections will have occurred from each end. This produces a complex deflection pattern. A function, **shkstrng**, was written to evaluate the series. Input parameters allow the user to select the number of series terms, the forcing frequency ($\omega \neq k\pi$) and the x and t limits. The MATLAB function **mesh** was employed to plot the surface $y(x, t)$ shown in Figure 2.3. The wave propagation mentioned above is clearly evident. A nearly resonant frequency $\omega = 1.95\pi$ was used. In Figure 2.4, we plot the deflection when the wave has moved halfway along the string. In Figure 2.5, we plot the deflection history at $x = 0.25$. That motion builds up from zero to fairly large amplitude by $t = 10$. A function called **motion** is also given which animates the string displacement. The animation is quite helpful for visualizing how waves travel along the string and reflect off the boundaries. Furthermore, progressive build up of the forced oscillations is evident. We leave these functions for the reader to study.

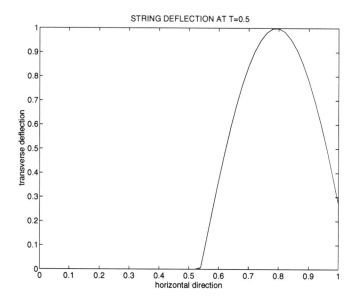

FIGURE 2.4. String Deflection at $T = 0.5$

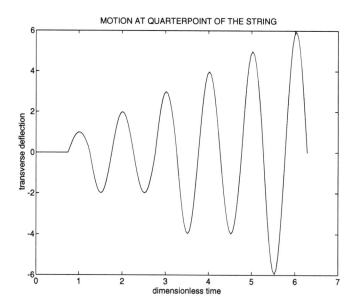

FIGURE 2.5. Motion at Quarterpoint of the String

MATLAB EXAMPLE

2.5.0.6 Script File stringmo

```
 1: % Example:  stringmo
 2: % ~~~~~~~~~~~~~~~~~~~
 3: % This is a driver program to illustrate motion
 4: % of a string having one end subjected to
 5: % harmonic oscillation.
 6: %
 7: % User m functions required:
 8: %    shkstrng, ploteasy
 9: %-------------------------------------------------
10:
11: fprintf('\nFORCED MOTION OF A VIBRATING ')
12: fprintf('STRING\n')
13: [y,t,x]=shkstrng(1.98*pi,150,0, ...
14:                  2*pi,151,0,1,51);
15: mesh(x,t,y), ylabel('time')
16: view([30,30]); xlabel('x axis');
17: zlabel('deflection');
18: title(['MOTION OF A STRING HAVING ONE END', ...
19:        ' SHAKEN HARMONICALLY'])
20: fprintf('\nPress RETURN to see deflection ')
21: fprintf('for t=0.5\n');
22: pause; genprint('strngsrf')
23: nt=sum(t<=.5); nx=sum(x<=.25);
24: ploteasy(x,y(nt,:),...
25:               'horizontal direction', ...
26:               'transverse deflection',...
27:               'STRING DEFLECTION AT T=0.5');
28: fprintf('\nPress RETURN to see deflection ')
29: fprintf('history at x=0.25\n')
30: pause, genprint('dflatep5')
31: ploteasy(t,y(:,nx), ...
32:               'dimensionless time',...
33:               'transverse deflection',...
34:               'MOTION AT QUARTERPOINT OF THE STRING')
35: genprint('motnqrtp')
36: fprintf('\nPress RETURN to see animation ')
37: fprintf('of the motion\n')
38: pause, motion(x,y), clf
39: disp(' '), disp('All Done')
```

2.5.0.7 Function shkstrng

```
 1: function [y,t,x]= ...
 2:           shkstrng(w,nsum,t1,t2,nt,x1,x2,nx)
 3: %
 4: % [y,t,x]=shkstrng(w,nsum,t1,t2,nt,x1,x2,nx)
 5: % ~~~~~~~~~~~~~~~~~~~~~~~~~~~~~~~~~~~~~~~~~~~~
 6: % Simulation of the motion of a string having
 7: % one end fixed and the other end shaken
 8: % harmonically.
 9: %
10: % w      - forcing frequency
11: % t1,t2 - minimum and maximum times
12: % nt     - number of time values
13: % x1,x2 - minimum and maximum x values
14: %           lying between zero and one
15: % nx     - number of x values
16: % t,x    - vectors of time and position values
17: % y      - matrix of transverse deflection
18: %           values having nt rows and nx
19: %           columns
20: %
21: % User m functions called:  none
22: %-----------------------------------------------
23:
24: t=t1+(t2-t1)/(nt-1)*(0:nt-1);
25: x=x1+(x2-x1)/(nx-1)*(0:nx-1)';
26: np=pi*(1:nsum); wsw=w/sin(w);
27: a=-(wsw./np) .*(sin(w-np)./(w-np) ...
28:    -sin(w+np)./(w+np));
29: y=(a(ones(nx,1),:).*sin(x*np))* ...
30:    sin(np(:)*t)+1/sin(w)*sin(w*x)*sin(w*t);
31: t=t(:); y=y';
```

2.5.0.8 Function ploteasy

```
 1: function ploteasy(x,y,xlabl,ylabl,titl)
 2: %
 3: % ploteasy(x,y,xlabl,ylabl,titl)
 4: % ~~~~~~~~~~~~~~~~~~~~~~~~~~~~~~~~
 5: % Easy plot function with a simple
 6: % argument list
```

```
 7: %
 8: % x,y   - data to be plotted
 9: % xlabl - horizontal axis label for the graph
10: % ylabl - vertical axis label for the graph
11: % titl  - title for the graph
12: %
13: % User m functions called:   none
14: %-----------------------------------------------
15:
16: plot(x,y)
17: if nargin==2, return, end
18: if nargin>2, xlabel(xlabl); end
19: if nargin>3, ylabel(ylabl); end
20: if nargin>4, title(titl); end
```

2.5.0.9   Function motion

```
 1: function motion(x,y,inct,trac)
 2: %
 3: % motion(x,y,inct,trac)
 4: % ~~~~~~~~~~~~~~~~~~~~~~
 5: % This function animates the motion history
 6: % of the string.
 7: %
 8: % x     - horizontal position coordinates
 9: %           corresponding to various columns
10: %           of matrix y
11: % y     - matrix with row j specifying the
12: %           string position at the j'th time
13: %           value
14: % inct  - the number of row increments used
15: %           to select positions for plotting.
16: %           Using inct=2 would plot every other
17: %           row of y. inct=1 is the default value.
18: % trac  - if this parameter is present,
19: %           successive plot images are left on
20: %           the screen. Otherwise, each
21: %           configuration is shown and removed
22: %           before the next image is shown. The
23: %           default choice is to remove
24: %           successive images.
25: %
26: % User m functions called:   none
```

```
27: %--------------------------------------------------
28:
29: if nargin ==2, inct=1; trac=0; end
30: if nargin ==3, trac=0; end
31: if inct > 1
32:   [nt,nx]=size(y); y=y(1:inct:nx,:);
33: end
34:
35: xmin=min(x); xmax=max(x);
36: ymin=min(y(:)); ymax=max(y(:)); clf
37: axis([xmin,xmax,2*ymin,2*ymax]);
38: [nt,nx]=size(y); axis off; hold on
39: for j=1:nt-1
40:    plot(x,y(j,:),'w'); drawnow;
41:    if trac ==0, cla; end
42: end
43: plot(x,y(nt,:),'w'); hold off
```

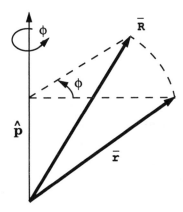

FIGURE 2.6. Rotation of a Vector About an Axis

2.6 Example on Animation of a Rotating Cube

Our final example shows the 3-dimensional motion of a rotating cube[1]. The object is described by tracing each of its edges. Assume that a cube of side length h initially has its faces parallel to the coordinate axes and its center at $(0,0,0)$. A rotation axis in the direction of a unit vector $\hat{p}$ is chosen and the y-axis of the cube is reoriented along the direction of $\hat{p}$. Then the cube center is moved to $(x_0, 0, 0)$. From that starting position on the x-axis, the cube is rotated about the y-axis, and positions after each $360/n$ degrees are shown graphically to animate the motion.

The mathematics needed to perform rotations is described as follows. Suppose a vector F is rotated through some angle ϕ about an axis $\hat{p}$ to produce a new vector $\bar{R}$ as shown in Figure 2.6. Simple geometrical calculations imply

$$\bar{R} = [1 - \cos(\phi)]\,\hat{p}(\hat{p} \cdot \bar{r}) + \cos(\phi)\bar{r} + \sin(\phi)\,[\hat{p} \times \bar{r}]$$

where $\hat{p}$ is a unit vector, $\hat{p} \cdot \bar{r}$ is a dot product, and $\hat{p} \times \bar{r}$ is a vector cross product. This vector equation can also be cast in matrix

---

[1]This example can also be implemented using the movie facility in MATLAB.

multiplication as

$$\begin{bmatrix} X_1 \\ X_2 \\ X_3 \end{bmatrix} = A \begin{bmatrix} x_1 \\ x_2 \\ x_3 \end{bmatrix}$$

where

$$A = (1 - \cos\phi) \begin{bmatrix} p_1^2 & p_1 p_2 & p_1 p_3 \\ p_2 p_1 & p_2^2 & p_2 p_3 \\ p_3 p_1 & p_3 p_2 & p_3^2 \end{bmatrix} + \cos\phi \begin{bmatrix} 1 & 0 & 0 \\ 0 & 1 & 0 \\ 0 & 0 & 1 \end{bmatrix} +$$

$$\sin\phi \begin{bmatrix} 0 & -p_2 & p_3 \\ p_1 & 0 & -p_3 \\ -p_1 & p_2 & 0 \end{bmatrix}$$

The transformation formula simplifies greatly when $\phi = \pi$ and can be written as

$$\bar{R} = 2\hat{p}(\hat{p} \cdot \bar{r}) - \bar{r}$$

This produces the well known Householder transformation [40] which can be conveniently employed to rotate any vector to assume some new direction chosen by the analyst. Assume a particular vector $\bar{r}$ is to be transformed to have a new direction, say $\bar{a}$. This can be achieved by using an axis which bisects the angles between $\bar{r}$ and $\bar{a}$, and then rotating 180° about that axis. Hence we let

$$\bar{p} = \bar{r}/\mathbf{norm}(\bar{r}) + \bar{a}/\mathbf{norm}(\bar{a}) \qquad \hat{p} = \bar{p}/\mathbf{norm}(p)$$

$$\bar{r}_{new} = 2\hat{p}(\hat{p} \cdot \bar{r}) - \bar{r}$$

These operations, which have obvious geometrical interpretation in three dimensions, generalize readily for vectors of arbitrary dimensionality and lead to an important method for reducing matrices to triangular form by successive rotations. (That method is applied to linear algebra problems in Golub and van Loan [40].)

Let us now discuss how to depict a polyhedral body with corners stored as columns of a matrix we will call U. When the body is rotated through an angle ϕ, the new corner coordinates contained in a matrix V are computed as $V = AU$, where matrix A is formulated as outlined above. In particular, if the object moves through a full revolution in m even increments, we take $\phi = 2\pi/m$

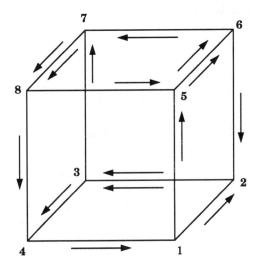

FIGURE 2.7. Point Indices for a Cube

and apply the transformation repeatedly. We can show a cube as a "wire frame" model by tracing out a sequence of corner coordinates which traverse all edges of the cube when the points are connected. A point sequence to accomplish this is illustrated below. Assume that the corner coordinates are stored as columns of U, with columns 1 to 4 identifying the bottom, and columns 5 to 8 identifying the top. According to Figure 2.7 we can traverse all edges of the cube by going from column to column in the matrix U(:,v) where v=[1:4,1,5:8,5,6,2,3,7,8,4].

A function was written to allow rotation of a cube having side length h and the center distance at $x0$ from the rotation axis. The cube can be positioned so an axis of symmetry makes a chosen angle with the vertically positioned rotation axis. The animated motion is achieved by writing an image on the screen and then erasing it before moving to the next position.

Several points about the program are worth noting. The animation is achieved by a looping process where we retrieve a column, reshape it into (x, y, z) coordinates, index it using vector v, plot the result, and move to the next position when the user presses a key. The speed of animation depends on how quickly the keyboard is pressed. Significant sections from the program are summarized below. Figure 2.8 was produced by allowing all rotated images to remain on the screen.

Routine	Line	Operation
boxrun	20-21	function **read** is employed to input the data.
	23-24	function **boxrot** is called to produce the animation.
boxrot	43-48	create box corner coordinates.
	50-57	rotate the cube axes and move center to a point on the x-axis.
	59-64	create matrix to perform rotation about the z-axis.
	66-75	compute different corner positions and save these points.
	85	set index vector to traverse all sides of the cube.
	87-100	show animated motion. Successive images can remain on the screen if desired.

ANIMATION OF A ROTATING CUBE

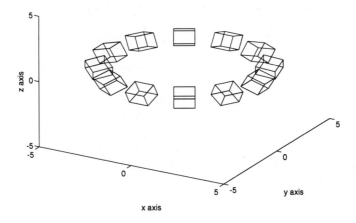

FIGURE 2.8. Various Positions for a Rotated Cube

MATLAB EXAMPLE

2.6.0.10 Function boxrun

```
 1: % Example:  boxrun
 2: % ~~~~~~~~~~~~~~~~~
 3: % Driver program to illustrate animated
 4: % rotation of a cube about a vertical axis
 5: %
 6: % User m functions required
 7: %    boxrot, read
 8: %-------------------------------------------------
 9:
10: fprintf('\nANIMATED ROTATION OF A CUBE ')
11: fprintf('ABOUT A VERTICAL AXIS\n\n')
12: fprintf('The graphics routine has the ')
13: fprintf('following parameters\n')
14: help boxrot
15: while 1
16:   fprintf('\nInput values for  p,  h, x0,')
17:   fprintf(' nstep, nloop, showall\n')
18:   fprintf('Try for example  pi/8, 1, 4,   ')
19:   fprintf('12,     1,     1    \n')
20:   [p,h,x0,nstep,nloop,showall]=read...
21:   ('Input  p,..., showall > ? ');
22:   if showall ==0
23:     [xmat,ymat,zmat]=boxrot(p,h,x0, ...
24:                             nstep,nloop);
25:   else
26:     [xmat,ymat,zmat]=boxrot(p,h,x0, ...
27:                             nstep,nloop,1);
28:   end
29:   v=input(['Enter 1 => stop or 2 => ' ...
30:             'continue > ? ']);
31:   if v==1, break; end
32: end
33: disp(' '), disp('All Done')
```

2.6.0.11 Function boxrot

```
 1: function [xmat,ymat,zmat]= ...
 2:          boxrot(p,h,x0,nstep,nloop,showall)
```

```
 3: %
 4: % [xmat,ymat,zmat]= ...
 5: %            boxrot(p,h,x0,nstep,nloop,showall)
 6: % ~~~~~~~~~~~~~~~~~~~~~~~~~~~~~~~~~~~~~~~~~~~~~~
 7: % This function applies rotation
 8: % transformations to animate the motion of a
 9: % cube rotating about a vertical axis (the z
10: % axis).
11: %
12: % Example use:
13: %
14: %          [xmat,ymat,zmat]=boxrot(pi/8,1,4,12,1,1)
15: %
16: % p          - angle between the cube axis
17: %              and the z axis
18: % h          - side length of the cube
19: % x0         - distance from the cube center
20: %              to the z axis
21: % nstep      - number of steps used to rotate
22: %              through 360 degrees (the
23: %              default is 80)
24: % nloop      - number of complete rotations
25: %              performed (the default is 3)
26: % showall    - if this parameter is present,
27: %              all images of the cube are left
28: %              showing on the screen.
29: %              Otherwise successive images are
30: %              erased.
31: % xmat,ymat, - coordinate arrays containing
32: % zmat         coordinates for successive
33: %              positions of the cube
34: %
35: % User m functions called:  none
36: %----------------------------------------------
37:
38: % Set defaults if some arguments are left out
39: if nargin==3, nstep=80; nloop=3; end
40: if nargin==4, nloop=3; end
41: if nargin<6, erase=1; else, erase=0; end
42:
43: % Create unit vectors along coordinate axes
44: i=eye(3,3); x=i(:,1); y=i(:,2); z=i(:,3);
45:
46: % Form corner coordinates of the cube
47: r=h/2*[x-y+z,x-y-z,-x-y-z,-x-y+z];
```

```
48: r=[r,r]; r(2,5:8)=-r(2,5:8);
49:
50: % Transform the x axis of the cube to be in
51: % the direction of the vector
52: % [cos(p); 0; -sin(p)].
53: q=[cos(p);0;-sin(p)]+[1;0;0];
54: q=q/norm(q); r=(2*q*q'-i)*r;
55:
56: % Shift the cube center to [x0; 0; 0]
57: r(1,:)=r(1,:)+x0*ones(1,8);
58:
59: % Create transformation matrix to rotate
60: % 360/nstep degrees at a time about the z axis
61: phi=2*pi/nstep; n=[0;0;1];
62: c=cos(phi); s=sin(phi);
63: a=(1-c)*n*n'+c*i+s*...
64:    [0,-n(3),n(2);n(3),0,-n(1);-n(2),n(1),0];
65:
66: % Create successive positions and save them in
67: % rows of xmat, ymat, zmat
68: xmat=zeros(nstep,8); ymat=xmat; zmat=xmat;
69: rnext=r;
70: xmat(1,:)=r(1,:); ymat(1,:)=r(2,:);
71: zmat(1,:)=r(3,:);
72: for j=2:nstep
73:    rnext=a*rnext; xmat(j,:)=rnext(1,:);
74:    ymat(j,:)=rnext(2,:); zmat(j,:)=rnext(3,:);
75: end
76:
77: % Determine window limits for plotting
78: dmin=floor(min([xmat(:);ymat(:);zmat(:)]));
79: dmax=ceil(max([xmat(:);ymat(:);zmat(:)]));
80: range=[dmin,dmax,dmin,dmax,dmin,dmax];
81:
82: % Form a vector of indices needed to plot all
83: % edges of the cube
84:
85: v=[1:4,1,5:8,5,6,2,3,7,8,4];
86:
87: % Draw the cube for various rotated positions
88: hold off; cla
89: xlabel('x axis'); ylabel('y axis');
90: zlabel('z axis');
91: title('ANIMATION OF A ROTATING CUBE');
92: axis(range); hold on; drawnow
```

```
 93:
 94: for k=1:nloop
 95:   for j=1:nstep
 96:     x=xmat(j,v); y=ymat(j,v); z=zmat(j,v);
 97:     plot3(x,y,z); view([30,30]); drawnow;
 98:     if ( erase & (k < nstep) ), cla, end
 99:   end
100: end
101: hold off
```

3

Summary of Concepts From Linear Algebra

3.1   Introduction

This chapter briefly reviews important concepts of linear algebra. We assume the reader already has some experience working with matrices, and linear algebra applied to solving simultaneous equations and eigenvalue problems. MATLAB has excellent capabilities to perform matrix operations using the fastest and most accurate algorithms currently available. The books by Strang [82] and Golub and van Loan [40] give comprehensive treatments of matrix theory and of algorithm developments accounting for effects of finite precision arithmetic. One beautiful aspect of matrix theory is that fairly difficult proofs often lead to remarkably simple results valuable to users not necessarily familiar with all of the theoretical developments. For instance, the property that every real symmetric matrix of order n has real eigenvalues and a set of n orthonormal eigenvectors can be understood and used by someone unfamiliar with the proof. The current chapter summarizes a number of fundamental matrix properties and some of the related MATLAB functions. These intrinsic functions are largely based on algorithms from the LINPACK and EISPACK software libraries [30, 36, 78]. Professor Cleve Moler actively contributed to development of these libraries and later he initiated development of an interactive computing environment which he named MATLAB. Readers should simultaneously study the current chapter and the MATLAB demo program on linear algebra.

3.2   Vectors, Norms, Linear Independence, and Rank

Consider an n by m matrix

$$A = [a_{ij}] \qquad 1 \le i \le n \qquad 1 \le j \le m$$

having real or complex elements. The shape of a matrix is computed by **size**(A) which returns a vector containing n and m. The matrix obtained by conjugating the matrix elements and interchanging columns and rows is called the transpose. Transposition is accomplished with a $'$ operator, so that

$$A\_\text{transpose} = A'$$

Transposition without conjugation of the elements can be performed as $A.'$ or as **conj**(A'). Of course, whenever A is real, A' is simply the traditional transpose.

The structure of a matrix A is characterized by the matrix rank and sets of basis vectors spanning four fundamental subspaces. The rank r is the maximum number of linearly independent rows or columns in the matrix. We discuss these spaces in the context of real matrices. The basic subspaces are:

1. The column space containing all vectors representable as a linear combination of the columns of A. The column space is also referred to as the range or the span.

2. The null space consisting of all vectors perpendicular to every row of A.

3. The row space consisting of all vectors which are linear combinations of the rows of A.

4. The left null space consisting of all vectors perpendicular to every column of A.

MATLAB has intrinsic functions to compute rank and subspace bases

- matrix_rank = **rank**(A)

- column_space = **orth**(A)

- null_space = **null**(A)

- row_space = **orth**$(A')'$

- left_null_space = **null**$(A')'$

The basis vectors produced by **null** and **orth** are orthonormal. They are generated using the QR algorithm [40]. Rank computation is performed by using the SVD [40] which gives high numerical accuracy but requires significantly more computation than competing methods. When computation time becomes important, it may be desirable to employ an alternative to SVD.

3.3   Systems of Linear Equations, Consistency, and Least Square Approximation

Let us discuss the problem of solving systems of simultaneous equations. Representing a vector B as a linear combination of the columns of A requires determination of a vector X to satisfy

$$AX = B \qquad \Longleftrightarrow \qquad \sum_{j=1}^{m} A(:,j)x(j) = B$$

where the j'th column of A is scaled by the j'th component of X to form the linear combination. The desired representation is possible if and only if B lies in the column space of A. This implies the consistency requirement that A and $[A, B]$ must have the same rank. Even when a system is consistent, the solution will not be unique unless all columns of A are independent. When matrix A, with n rows and m columns, has rank r less than m, the general solution of $AX = B$ is expressible as any particular solution plus an arbitrary linear combination of $m - r$ vectors forming a basis for the null space. MATLAB gives the solution vector as $X = A \backslash B$. When r is less than m, MATLAB produces a least square solution having as many components as possible set equal to zero. Furthermore, MATLAB outputs a short warning message indicating rank deficiency.

In instances where the system is inconsistent, regardless of how X is chosen, the error vector defined by

$$E = AX - B$$

can never be zero. An approximate solution can be obtained by making E normal to the columns of A. We get

$$A'AX = A'B$$

which is known as the system of normal equations. They are also referred to as least square error equations. It is not difficult to show that the same equations result by requiring E to have minimum length. The normal equations are always consistent and are uniquely solvable when $\mathbf{rank}(A) = m$. A comprehensive discussion of least square approximation and methods for solving overdetermined systems is presented by Lawson and Hanson [52]. It is instructive to examine the results obtained from the normal equations when A is square and nonsingular. The least square solution would give

$$X = (A'A)^{-1}A'B = A^{-1}(A')^{-1}A'B = A^{-1}B$$

Therefore, the least square solution simply reduces to the exact solution of $AX = B$ for a consistent system. MATLAB handles both consistent and inconsistent systems as $X = A\backslash B$. However, it is only sensible to use the least square solution of an inconsistent system when AX produces an acceptable approximation to B. This implies

$$\mathbf{norm}(AX - B) < tol * \mathbf{norm}(B)$$

where tol is suitably small.

A simple but important application of overdetermined systems arises in curve fitting. An equation of the form

$$y(x) = \sum_{j=1}^{m} f_j(x)c_j$$

involving known functions $f_j(x)$, such as x^{j-1} for polynomials, must approximately match data values (X_i, Y_i), $1 \leq i \leq n$, with $n > m$. We simply write an overdetermined system

$$\sum_{j=1}^{n} f_j(X_i)c_j \approx Y_i \qquad 1 \leq i \leq n$$

and obtain the least square solution. The approximation is acceptable if the error components

$$e_i = \sum_{j=1}^{m} f_j(X_i)c_j - Y_i$$

are small enough and the function $y(x)$ is also acceptably smooth between the data points.

Let us illustrate how well MATLAB handles simultaneous equations by constructing the steady-state solution of the matrix differential equation

$$M\ddot{x} + C\dot{x} + Kx = F_1 \cos(\omega t) + F_2 \sin(\omega t)$$

where M, C, and K are constant matrices and F_1 and F_2 are constant vectors. The steady state solution has the form

$$x = X_1 \cos(\omega t) + X_2 \sin(\omega t)$$

where X_1 and X_2 are chosen so that the differential equation is satisfied. Evidently

$$\dot{x} = -\omega X_1 \sin(\omega t) + \omega X_2 \cos(\omega t)$$

and

$$\ddot{x} = -\omega^2 x$$

Substituting the assumed form into the differential equation and comparing sine and cosine terms on both sides yields

$$(K - \omega^2 M)X_1 + \omega C X_2 = F_1$$

$$-\omega C X_1 + (K - \omega^2 M)X_2 = F_2$$

The equivalent partitioned matrix is

$$\left[\begin{array}{c|c} (K - \omega^2 M) & \omega C \\ \hline -\omega C & (K - \omega^2 M) \end{array} \right] \left[\begin{array}{c} X_1 \\ X_2 \end{array} \right] = \left[\begin{array}{c} F_1 \\ F_2 \end{array} \right]$$

A simple MATLAB function to produce $X1$ and $X2$ when M, C, K, $F1$, and $F2$ are known is

```
function [x1,x2,xmax]=forcresp(m,c,k,f1,f2,w)
kwm=k-(w*w)*m; wc=w*c;
x=[kwm,wc;-wc,kwm]\[f1;f2]; n=length(f1);
x1=x(1:n); x2=x(n+1:2*n);
xmax=sqrt(x1.*x1+x2.*x2);
```

The vector, **xmax**, defined in the last line of the function above, has components specifying the maximum amplitude of each component of the steady state solution. The main computation in this function occurs in the third line, where matrix concatenation is employed to form a system of $2n$ equations with x being the concatenation of X_1 and X_2. The last line uses vector indexing to extract X_1 and X_2 from x. The notational simplicity of MATLAB is elegantly illustrated by these features: a) any required temporary storage is assigned and released dynamically, b) no looping operations are needed, c) matrix concatenation and inversion are accomplished with intrinsic functions using matrices and vectors as sub-elements of other matrices, d) and extraction of sub-vectors is accomplished by use of vector indices. The important idea communicated by this function is that mathematical notation and MATLAB language constructs are essentially identical in this important application.

3.4   Applications of Least Square Approximation

The idea of solving an inconsistent system of equations in the least square sense, so that some required condition is approximately satisfied, has numerous applications. Typically, we are dealing with a large number of equations (several hundred is common) involving a much smaller number of parameters used to closely fit some constraint. Linear boundary value problems often require the solution of a differential equation applicable in the interior of a region while the function values are known on the boundary. This type of problem can sometimes be handled by using a series of functions which satisfy the differential equation exactly. Weighting the component solutions to approximately match the remaining boundary condition may lead to useful results. Below, we examine three instances where least square approximation is helpful.

3.4.1   A Membrane Deflection Problem

Let us illustrate how least square approximation can be used to compute the transverse deflection of a membrane subjected to uniform pressure. The transverse deflection u for a membrane which has zero deflection on a boundary L satisfies the differential equa-

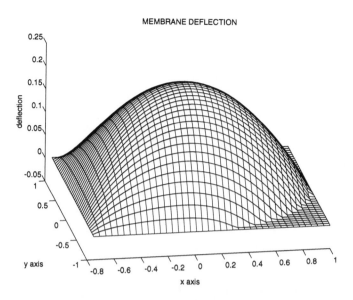

FIGURE 3.1. Surface Plot of Membrane

tion

$$\frac{\partial^2 u}{\partial x^2} + \frac{\partial^2 u}{\partial y^2} = -\gamma \qquad \text{(x,y) inside L}$$

where γ is a physical constant. Properties of harmonic functions [16] imply that the differential equation is satisfied by a series of the form

$$u = \gamma[\frac{-|z|^2}{4} + \sum_{j=0}^{n} c_j \mathbf{real}(z^{j-1})]$$

where $z = x + \imath y$ and constants c_j are chosen to make the boundary deflection as small as possible, in the least square sense. As a specific example, we analyze a membrane consisting of a rectangular part on the left joined with a semicircular part on the right. The surface plot in Figure 3.1 and the contour plot in Figure 3.2 were produced by the function **membrane** listed below. This function generates boundary data, solves for the series coefficients, and constructs plots depicting the deflection pattern. The results obtained using a twenty-term series satisfy the boundary conditions quite well.

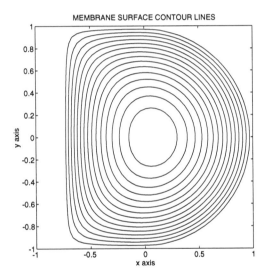

FIGURE 3.2. Membrane Surface Contour Lines

MATLAB EXAMPLE

3.4.1.1   Function membrane

```
1: function [dfl,cof]=membrane(h,np,ns,nx,ny)
2: %
3: % [dfl,cof]=membrane(h,np,ns,nx,ny)
4: % ~~~~~~~~~~~~~~~~~~~~~~~~~~~~~~~~~~
5: % This function computes the transverse
6: % deflection of a uniformly tensioned membrane
7: % which is subjected to uniform pressure. The
8: % membrane shape is a rectangle of width h and
9: % height two joined with a semicircle of
10: % diameter two.
11: %
12: % Example use:  membrane(0.75,100,50,40,40);
13: %
14: % h        - the width of the rectangular part
15: % np       - the number of least square points
16: %            used to match the boundary
17: %            conditions in the least square
18: %            sense is about 3.5*np
19: % ns       - the number of terms used in the
20: %            approximating series evaluate
21: %            deflections. The series has the
22: %            form
23: %
24: %            dfl = abs(z)^2/4 +
25: %                  sum({j=1:ns},cof(j)*
26: %                  real(z^(j-1)))
27: %
28: % nx,ny    - the number of x points and y points
29: %            used to compute deflection values
30: %            on a rectangular grid
31: % dfl      - computed array of deflection values
32: % cof      - coefficients in the series
33: %            approximation
34: %
35: % User m functions called:  genprint
36: %-------------------------------------------------
37:
38: % Generate boundary points for least square
39: % approximation
40: z=[exp(i*linspace(0,pi/2,round(1.5*np))),...
```

```
41:    linspace(i,-h+i,np),...
42:    linspace(-h+i,-h,round(np/2))];
43: z=z(:);
44:
45: % Form the least square equations and solve
46: % for series coefficients
47: a=ones(length(z),ns);
48: for j=2:ns, a(:,j)=a(:,j-1).*z; end
49: cof=real(a)\(z.*conj(z))/4;
50:
51: % Generate a rectangular grid for evaluation
52: % of deflections
53: xv=linspace(-h,1,nx); yv=linspace(-1,1,ny);
54: [x,y]=meshgrid(xv,yv); z=x+i*y;
55:
56: % Evaluate the deflection series on the grid
57: dfl=-z.*conj(z)/4+ ...
58:     real(polyval(cof(ns:-1:1),z));
59:
60: % Set values outside the physical region of
61: % interest to zero
62: dfl=real(dfl).*(1-((abs(z)>=1)&(real(z)>=0)));
63:
64: % Make surface and contour plots
65: mesh(x,y,dfl); view(-10,30);
66: xlabel('x axis'); ylabel('y axis');
67: zlabel('deflection')
68: title('MEMBRANE DEFLECTION'); disp(' ')
69: disp('Press RETURN to show a contour plot')
70: pause; genprint('membdefl');
71: contour(x,y,dfl,15);
72: axis([-1,1,-1,1]); axis('square')
73: xlabel('x axis'); ylabel('y axis')
74: title('MEMBRANE SURFACE CONTOUR LINES')
75: genprint('membcntr');
```

3.4.2   MIXED BOUNDARY VALUE PROBLEM FOR A FUNCTION HARMONIC INSIDE A CIRCULAR DISK

Problems where a partial differential equation is to be solved inside a region with certain conditions imposed on the boundary occur in many situations. Often the differential equation is solvable exactly in a series form containing arbitrary linear combinations of known functions. An approximation procedure imposing the boundary conditions to compute the series coefficients produces a satisfactory solution if the desired boundary conditions are found to be well satisfied. Consider a mixed boundary value problem in potential theory [63] pertaining to a circular disk of unit radius. We seek $u(r, \theta)$ where function values are specified on one part of the boundary and normal derivative values are specified on the remaining part. The mathematical formulation is

$$\frac{\partial^2 u}{\partial r^2} + \frac{1}{r}\frac{\partial u}{\partial r} + \frac{1}{r^2}\frac{\partial^2 u}{\partial \theta^2} = 0 \qquad 0 \leq r < 1 \qquad 0 \leq \theta \leq 2\pi$$

$$u(1, \theta) = f(\theta) \qquad -\alpha < \theta < \alpha$$

$$\frac{\partial u}{\partial r}(1, \theta) = g(\theta) \qquad \alpha < \theta < 2\pi - \alpha$$

The differential equation has a series solution of the form

$$u(r, \theta) = c_0 + \sum_{n=1}^{\infty} r^n[c_n \cos(n\theta) + d_n \sin(n\theta)],$$

the boundary conditions require

$$c_0 + \sum_{n=1}^{\infty}[c_n \cos(n\theta) + d_n \sin(n\theta)] = f(\theta) \qquad -\alpha < \theta < \alpha,$$

and

$$\sum_{n=1}^{\infty} n[c_n \cos(n\theta) + d_n \sin(n\theta)] = g(\theta) \qquad \alpha < \theta < 2\pi - \alpha$$

The series coefficients can be obtained by least square approximation. Let us explore the utility of this approach by considering a particular problem for a field which is symmetric about the x-axis. We want to solve

$$\nabla^2 u = 0 \qquad r < 1$$

$$u(1,\theta) = \cos(\theta) \qquad |\theta| < \pi/2$$

$$\frac{\partial u}{\partial r}(1,\theta) = 0 \qquad \pi/2 < |\theta| \leq \pi$$

This problem characterizes steady state heat conduction in a cylinder with the left half insulated and the right half held at a known temperature. The appropriate series solution is

$$u = \sum_{n=0}^{\infty} c_n r^n \cos(n\theta)$$

subject to

$$\sum_{n=0}^{\infty} c_n \cos(n\theta) = \cos(\theta) \qquad |\theta| < \pi/2$$

$$\sum_{n=0}^{\infty} n c_n \cos(n\theta) = 0 \qquad \pi/2 < |\theta| \leq \pi$$

We solve the problem by truncating the series after a hundred or so terms and forming an overdetermined system derived by imposition of both boundary conditions. The success of this procedure depends on the series converging rapidly enough so that a system of least square equations having reasonable order and satisfactory numerical condition results. It can be shown by complex variable methods (see Muskhelishvili [63]) that the exact solution of our problem is given by

$$u = \mathbf{real}\left(z + z^{-1} + (1 - z^{-1})\sqrt{z^2 + 1}\right)/2 \qquad |z| \leq 1$$

where the square root is defined for a branch cut along the right half of the unit circle with the chosen branch being that which equals $+1$ at $z = 0$. Readers familiar with analytic function theory can verify that the boundary values of u yield

$$u(1,\theta) = \cos(\theta) \qquad |\theta| \leq \pi/2$$

$$u(1,\theta) = \cos(\theta) + \sin(|\theta|/2)\sqrt{2|\cos(\theta)|} \qquad \pi/2 \leq |\theta| \leq \pi$$

A least square solution is presented in function **mbvp**. Results from a series of 100 terms are shown in Figure 3.3. The series solution is accurate within about one percent error except for points near $\theta = \pi/2$. Although the results are not shown here, using 300 terms gives a solution error which is almost imperceptible on a graph. Hence the least square series solution provides a reasonable method to handle the mixed boundary value problem.

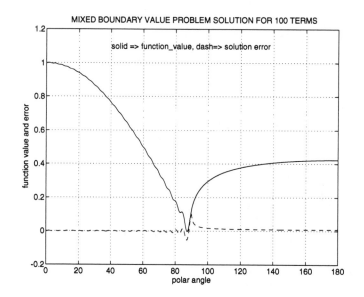

FIGURE 3.3. Mixed Boundary Value Problem Solution

MATLAB EXAMPLE

3.4.2.1 Script file mbvprun

```
 1: % Example:  mbvprun
 2: % ~~~~~~~~~~~~~~~~~~
 3: % Mixed boundary value problem for a function
 4: % harmonic inside a circle.
 5: %
 6: % User m functions required:
 7: %    mbvp
 8: %---------------------------------------------
 9:
10: disp('Calculating');
11:
12: % Set data for series term and boundary
13: % condition points
14: nser=100; nf=200; ng=200; neval=500;
15:
16: % Compute the series coefficients
17: [cof,y]=mbvp('cos',pi/2,nser,nf,ng,neval);
18:
19: % Evaluate the exact solution for comparision
20: thp=linspace(0,pi,neval)';
21: y=cos(thp*(0:nser-1))*cof;
22: ye=cos(thp)+sin(thp/2).* ...
23:    sqrt(2*abs(cos(thp))).*(thp>=pi/2);
24:
25: % Plot results showing the accuracy of the
26: % least square solution
27: thp=thp*180/pi; plot(thp,y,'w',thp,y-ye,'w--');
28: xlabel('polar angle');
29: ylabel('function value and error')
30: title(['MIXED BOUNDARY VALUE PROBLEM ', ...
31:       'SOLUTION FOR ',int2str(nser),' TERMS']);
32: text(40,1.1,['solid => function_value, ', ...
33:               'dash=> solution error']);
34: grid
35: genprint('mbvp')
```

3.4.2.2   Function mbvp

```
 1: function [cof,y]= ...
 2:           mbvp(func,alp,nser,nf,ng,neval)
 3: %
 4: % [cof,y]=mbvp(func,alp,nser,nf,ng,neval)
 5: % ~~~~~~~~~~~~~~~~~~~~~~~~~~~~~~~~~~~~~~~~~
 6: % This function solves approximately a mixed
 7: % boundary value problem for a function which
 8: % is harmonic inside the unit disk, symmetric
 9: % about the x axis, and has boundary conditions
10: % involving function values on one part of the
11: % boundary and zero gradient elsewhere.
12: %
13: % func       - function specifying the function
14: %               value between zero and alp
15: %               radians
16: % alp        - angle between zero and pi which
17: %               specifies the point where
18: %               boundary conditions change from
19: %               function value to zero gradient
20: % nser       - number of series terms used
21: % nf         - number of function values
22: %               specified from zero to alp
23: % ng         - number of points from alp to pi
24: %               where zero normal derivative is
25: %               specified
26: % neval      - number of boundary points where
27: %               the solution is evaluated
28: % cof        - coefficients in the series
29: %               solution
30: % y          - function values for the solution
31: %
32: % User m functions called:  none.
33: %------------------------------------------------
34:
35: % Create evenly spaced points to impose
36: % boundary conditions
37: th1=linspace(0,alp,nf);
38: th2=linspace(alp,pi,ng+1); th2(1)=[];
39:
40: % Form an overdetermined system based on the
41: % boundary conditions
42: yv=feval(func,th1);
```

```
43: cmat=cos([th1(:);th2(:)]*(0:nser-1));
44: [nr,nc]=size(cmat);
45: cmat(nf+1:nr,:)=...
46:   (ones(ng,1)*(0:nser-1)).*cmat(nf+1:nr,:);
47: cof=cmat\[yv(:);zeros(ng,1)];
48:
49: % Evaluate the solution on the boundary
50: thp=linspace(0,pi,neval)';
51: y=cos(thp*(0:nser-1))*cof;
```

3.4.3 Using Rational Functions to Conformally Map a Circular Disk Onto a Square

Another problem illustrating the value of least square approximation arises in connection with an example discussed earlier in Section 2.4 where a slowly convergent power series was used to map the interior of a circle onto the interior of a square [64]. It is sometimes possible for slowly convergent power series of the form

$$w = f(z) = \sum_{j=0}^{N} c_j z^j \qquad |z| \leq 1$$

to be replaceable by a rational function

$$w = \frac{\displaystyle\sum_{j=0}^{n} a_j z^j}{1 + \displaystyle\sum_{j=1}^{m} b_j z^j}$$

Of course, the polynomial is simply a special rational function form with $m = 0$ and $n = N$. This rational function implies

$$\sum_{j=0}^{n} a_j z^j - w \sum_{j=1}^{m} b_j z^j = w$$

Coefficients a_j and b_j can be computed by forming least square equations based on boundary data. In some cases, the resulting equations are rank deficient and it is safer to solve a system of the form $UY = V$ as $Y = \mathbf{pinv}(U) * V$ rather than using $Y = U \backslash V$. The former solution uses the pseudo inverse function **pinv** which automatically sets to zero any solution components that are undetermined.

Two functions **ratcof** and **raterp** were written to compute rational function coefficients and to evaluate the rational function for general matrix arguments. These functions are useful to examine the conformal mapping of the circular disk $|z| \leq 1$ onto the square defined by $|\mathbf{real}(w)| \leq 1$, $|\mathbf{imag}(w)| \leq 1$. A polynomial approximation of the mapping function has the form

$$w/z = \sum_{j=0}^{N} c_j (z^4)^j$$

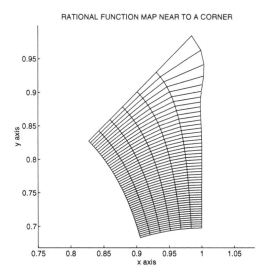

FIGURE 3.4. Rational Function Map Near to Corner

where N must be quite large in order to avoid excessive corner rounding. If we evaluate w versus z on the boundary for large N (500 or more), and then develop a rational function fit with $n = m = 10$, a reasonably good representation of the square results without requiring a large number of series terms. The following program illustrates the use of functions **ratcof** and **raterp**. It also includes a function **sqmp** to generate coefficients in the Schwarz-Christoffel series and a function **cplot** to produce an undistorted grid plot corresponding to a matrix of complex numbers. Figure 3.4 shows the geometry mapping produced in the vicinity of a corner.

MATLAB EXAMPLE

3.4.3.1   Script file makratsq

```
 1: % Example:  makratsq
 2: % ~~~~~~~~~~~~~~~~~~
 3: % Create a rational function map of a unit disk
 4: % onto a square.
 5: %
 6: % User m functions required:
 7: %    sqmp, ratcof, raterp, cplot
 8: %-------------------------------------------------
 9:
10: disp(' ');
11: disp('RATIONAL FUNCTION MAPPING OF A CIRCULAR')
12: disp('         DISK ONTO A SQUARE'); disp(' ')
13: disp('Calculating'); disp(' ')
14:
15: % Generate boundary points given by the
16: % Schwarz-Christoffel transformation
17: nsc=501; np=401; ntop=10; nbot=10;
18: z=exp(i*linspace(0,pi/4,np));
19: w=sqmp(nsc,1,1,1,0,45,np);
20: w=mean(real(w))+i*imag(w);
21: z=[z,conj(z)]; w=[w,conj(w)];
22:
23: % Compute the series coefficients for a
24: % rational function fit to the boundary data
25: [top,bot]=ratcof(z.^4,w./z,ntop,nbot);
26: top=real(top); bot=real(bot);
27:
28: % The above calculations produce the following
29: % coefficients
30: % [top,bot]=
31: %          1.0787    1.4948
32: %          1.5045    0.1406
33: %          0.0353   -0.1594
34: %         -0.1458    0.1751
35: %          0.1910   -0.1513
36: %         -0.1797    0.0253
37: %          0.0489    0.2516
38: %          0.2595    0.1069
39: %          0.0945    0.0102
40: %          0.0068    0.0001
```

```
41:
42: % Generate a polar coordinate grid to describe
43: % the mapping near the corner of the square.
44: % Then evaluate the mapping function.
45: r1=.95; r2=1; nr=7; t1=.9*pi/4; t2=pi/4; nt=51;
46: [r,th]= ...
47:         meshdom(linspace(r1,r2,nr), ...
48:         linspace(t1,t2,nt));
49: z=r.*exp(i*th); w=z.*raterp(top,bot,z.^4);
50:
51: % Plot the mapped geometry
52: cplot(w);
53: title('RATIONAL FUNCTION MAP NEAR TO A CORNER')
54: xlabel('x axis'); ylabel('y axis')
55: genprint('ratsqmap')
```

3.4.3.2   Function sqmp

```
 1: function [w,b]=sqmp(m,r1,r2,nr,t1,t2,nt)
 2: %
 3: % [w,b]=sqmp(m,r1,r2,nr,t1,t2,nt)
 4: % ~~~~~~~~~~~~~~~~~~~~~~~~~~~~~~~~~
 5: % This function evaluates the conformal
 6: % mapping produced by the Schwarz-Christoffel
 7: % transformation w(z) mapping abs(z)<=1 inside
 8: % a square having a side length of two.  The
 9: % transformation is approximated in series form
10: % which converges very slowly near the corners.
11: %
12: % m        - number of series terms used
13: % r1,r2,nr - abs(z) varies from r1 to r2 in
14: %            nr steps
15: % t1,t2,nt - arg(z) varies from t1 to t2 in
16: %            nt steps (t1 and t2 are
17: %            measured in degrees)
18: % w        - points approximating the square
19: % b        - coefficients in the truncated
20: %            series expansion which has
21: %            the form
22: %
23: %            w(z)=sum({j=1:m},b(j)*z*(4*j-3))
24: %
25: % User m functions called:  none.
```

```
26: %--------------------------------------------------
27:
28: % Generate polar coordinate grid points for the
29: % map. Function linspace generates vectors with
30: % equally spaced components.
31: r=linspace(r1,r2,nr)';
32: t=pi/180*linspace(t1,t2,nt);
33: z=(r*ones(1,nt)).*(ones(nr,1)*exp(i*t));
34:
35: % Compute the series coefficients and evaluate
36: % the series
37: k=1:m-1;
38: b=cumprod([1,-(k-.75).*(k-.5)./(k.*(k+.25))]);
39: b=b/sum(b);
40: w=z.*polyval(b(m:-1:1),z.^4);
```

3.4.3.3   Function ratcof

```
 1: function [a,b]=ratcof(xdata,ydata,ntop,nbot)
 2: %
 3: % [a,b]=ratcof(xdata,ydata,ntop,nbot)
 4: % ~~~~~~~~~~~~~~~~~~~~~~~~~~~~~~~~~~~~~
 5: % Determine a and b to approximate ydata as a
 6: % rational function of the variable xdata. The
 7: % function has the form:
 8: %
 9: %         y(x) = sum(1=>ntop)( a(j)*x^(j-1) ) /
10: %            ( 1 + sum(1=>nbot)( b(j)*x^(j)) )
11: %
12: % xdata,ydata - input data vectors
13: %               (real or complex)
14: % ntop,nbot   - number of series terms used in
15: %               the numerator and the
16: %               denominator.
17: %
18: % User m functions called:  none.
19: %--------------------------------------------------
20:
21: ydata=ydata(:);   xdata=xdata(:);
22: m=length(ydata);
23: if nargin==3, nbot=ntop; end;
24: x=ones(m,ntop+nbot);
25: x(:,ntop+1)=-ydata.*xdata;
```

84

```
26: for i=2:ntop, ...
27:   x(:,i)=xdata.*x(:,i-1); end
28: for i=2:nbot, ...
29:   x(:,i+ntop)=xdata.*x(:,i+ntop-1); end
30: ab=pinv(x)*ydata;
31: a=ab(1:ntop);
32: b=ab(ntop+1:ntop+nbot);
```

3.4.3.4 Function raterp

```
 1: function y=raterp(a,b,x)
 2: %
 3: % y=raterp(a,b,x)
 4: % ~~~~~~~~~~~~~~~~
 5: % This function interpolates using coefficients
 6: % from function ratcof.
 7: %
 8: % a,b - polynomial coefficients from function
 9: %        ratcof
10: % x   - argument at which function is evaluated
11: % y   - computed rational function values
12: %
13: % User m functions called:  none.
14: %-----------------------------------------------
15:
16: a=flipud(a(:)); b=flipud(b(:));
17: y=polyval(a,x)./(1+x.*polyval(b,x));
```

3.4.3.5 Function cplot

```
 1: function cplot(z,wsiz)
 2: %
 3: % cplot(z,wsiz)
 4: % ~~~~~~~~~~~~~
 5: % Plot complex array z undistorted in a square
 6: % window. If the vector wsiz is present, it is
 7: % used to set the window size.  Otherwise, the
 8: % z array is examined to set the window.
 9: %
10: % z   - complex vector used to plot imag(z)
11: %          versus real z
```

```
12: % wsiz - vector to set window limits (optional)
13: %
14: % User m functions called:  none.
15: %-----------------------------------------------
16:
17: [nrow,ncol]=size(z);
18: if nargin==1,  rl=real(z(:)); img=imag(z(:));
19:    xmin=min(rl); xmax=max(rl);
20:    ymin=min(img); ymax=max(img);
21: else
22:    xmin=wsiz(1); xmax=wsiz(2);
23:    ymin=wsiz(3); ymax=wsiz(4);
24: end
25: w=max((xmax-xmin),(ymax-ymin))*.55;
26: xc=.5*(xmin+xmax); yc=.5*(ymin+ymax);
27: v=[xc-w,xc+w,yc-w,yc+w];
28: axis(v), axis('square')
29: if ncol>1
30:    for i=1:nrow,...
31:      plot(real(z(i,:)),imag(z(i,:)),'-'),...
32:      hold on, end;
33: end
34: if nrow>1
35:    for j=1:ncol,...
36:      plot(real(z(:,j)),imag(z(:,j)),'-'),...
37:      hold on, end;
38: end
39: hold off; axis('normal');
```

3.5 Eigenvalue Problems

3.5.1 STATEMENT OF THE PROBLEM

Another important linear algebra problem involves computation of nonzero vectors X and numbers λ such that

$$AX = \lambda X$$

where A is a square matrix of order n. The number λ, which can be real or complex, is called an eigenvalue corresponding to an eigenvector X. The eigenvalue equation implies

$$[I\lambda - A]X = 0$$

so that λ values must be selected to make $I\lambda - A$ singular. The polynomial

$$f(\lambda) = \det(I\lambda - A) = \lambda^n + c_1\lambda^{n-1} + \ldots + c_n\lambda^0$$

is called the characteristic equation and its roots are the eigenvalues. It can be factored into

$$f(\lambda) = (\lambda - \lambda_1)(\lambda - \lambda_2)\cdots(\lambda - \lambda_n)$$

The eigenvalues are generally complex numbers and some of the roots may be repeated. In the usual situation, distinct roots $\lambda_1, \cdots, \lambda_n$ yield n linearly independent eigenvectors obtained by solving

$$(A - \lambda_j I)X_j = 0 \qquad 1 \leq j \leq n$$

The case involving repeated eigenvalues is more complicated. Suppose a particular eigenvalue such as λ_1 has multiplicity k, then the general solution of

$$(A - \lambda_1 I)X = 0$$

will yield as few as one, or as many as k, linearly independent vectors. If fewer than k independent eigenvectors are found for any root of multiplicity k, then matrix A is called defective. Occurrence of a defective matrix is not typical. It usually implies special behavior of the associated physical system. In the general situation the complete set of eigenvectors can be written as

$$A[X_1, \cdots, X_n] = [X_1\lambda_1, \cdots, X_n\lambda_n]$$

$$= [X_1, \cdots, X_n]\mathbf{diag}(\lambda_1, \cdots, \lambda_n)$$

or

$$AU = U\Lambda$$

where U contains the eigenvectors as columns and Λ is a diagonal matrix with eigenvalues on the diagonal. When the eigenvalues are linearly independent, the matrix U, which is known as the modal matrix, is nonsingular. This allows A to be expressed as

$$A = U\Lambda U^{-1}$$

which is convenient for various computational purposes. Unfortunately, this decomposition is not always possible. It does exist whenever the eigenvalues are distinct and for special types of matrices such as symmetric matrices which are discussed below.

A matrix A is symmetric if $A = A'$ where A' is obtained by interchanging columns and rows, and conjugating all elements. It can be shown that symmetric matrices have eigenvalues which are all real and a linearly independent set of eigenvectors. Furthermore, the eigenvectors X_j and X_k for any two unequal eigenvalues turn out to satisfy an orthogonality condition

$$X_j' X_k = 0 \qquad j \neq k$$

Eigenvectors belonging to the same repeated eigenvalue are not automatically orthogonal. Nevertheless, they can be replaced by an equivalent orthogonal set by applying a process called Gram-Schmidt orthogonalization [40]. In cases of interest here, the symmetric matrix A can be assumed to have real elements. Therefore the eigenvalues are real with eigenvectors satisfying $X_i' X_j = \delta_{ij}$, where δ_{ij} is the Kronecker delta symbol. The orthogonality condition is equivalent to the statement that $U'U = I$, so a real symmetric matrix can be expressed as

$$A = U\Lambda U'$$

It is important in MATLAB that the symmetry condition $A' = A$ be satisfied perfectly. A matrix which is symmetric, except for roundoff, can be replaced by $(A + A')/2$ to guarantee perfect symmetry. The MATLAB function **eig** computes eigenvalues and eigenvectors. When a matrix is perfectly symmetric (not just to machine accuracy), **eig** generates real eigenvalues and orthonormalized eigenvectors.

An important property of symmetric matrices and the related orthonormal eigenvector set occurs in connection with quadratic forms expressed as

$$F(Y) = Y'AY$$

where Y is an arbitrary real vector and A is real symmetric. The function $F(Y)$ is a one-by-one matrix, hence it is a scalar function. The algebraic sign of the form for arbitrary nonzero choices of Y is important in physical applications. Let us use the eigenvector decomposition of A to write

$$F = Y'U'\Lambda UY = (UY)'\Lambda(UY)$$

Taking $X = UY$ and $Y = U'X$ gives

$$F = X'\Lambda X = \lambda_1 x_1^2 + \lambda_2 x_2^2 + \lambda_3 x_3^2 + \ldots + \lambda_n x_n^2$$

This diagonal form makes the algebraic character of F evident. If all λ_i are positive, then F is evidently positive whenever X has at least one nonzero component. Then the quadratic form is called positive definite. If the eigenvalues are all positive or zero, the form is called positive semidefinite since the form cannot assume a negative value but can equal zero without having $X = 0$. When both negative and positive eigenvalues are present, the form can change sign and is termed indefinite. When the eigenvalues are all negative, the form is classified as negative definite. Perhaps the most important of these properties is that a necessary and sufficient condition for the form to be positive definite is that all eigenvalues of A be positive.

An important generalization of the standard eigenvalue problem has the form

$$AX = \lambda BX$$

for arbitrary A and nonsingular B. If B is well conditioned, then it is computationally attractive to simply solve

$$B^{-1}AX = \lambda X$$

In general, it is safer, but much more time consuming, to call **eig** as

```
[EIGVECS,EIGVALS]=eig(A,B)
```

This returns the eigenvectors as columns of **EIGVECS** and also gives a diagonal matrix **EIGVALS** which contains the eigenvalues.

3.5.2 APPLICATION TO SOLUTION OF MATRIX DIFFERENTIAL EQUATIONS

One of the most important applications of eigenvalues concerns the solution of the linear, constant-coefficient matrix differential equation

$$B\dot{Y}(t) = AY(t) \qquad Y(0) = Y_0$$

Component solutions can be written as

$$Y = Xe^{\lambda t} \qquad \dot{Y} = \lambda Xe^{\lambda t}$$

where X and λ are constant. Substitution into the differential equation gives

$$(A - \lambda B)Xe^{\lambda t} = 0$$

Since $e^{\lambda t}$ cannot vanish we need

$$AX = \lambda BX$$

After the eigenvalues and eigenvectors have been computed, a general solution is constructed as a linear combination of component solutions

$$Y = \sum_{j=1}^{n} X_j e^{\lambda_j t} c_j$$

The constants c_j are obtained by imposing the initial condition

$$Y(0) = [X_1, X_2, \dots, X_n]c$$

Assuming that the eigenvectors are linearly independent we get

$$c = [X_1, \dots, X_n]^{-1} Y_0$$

3.6 Column Space, Null Space, Orthonormal Bases, and SVD

A final topic discussed in this chapter is the factorization known as singular value decomposition, or SVD. We will briefly explain the structure of SVD and some of its applications. It is known that any real matrix having n rows, m columns, and rank r can be decomposed into the form

$$A = USV'$$

where

- U is an orthogonal n by n matrix such that $U'U = I$

- V is an orthogonal m by m matrix such that $V'V = I$

- S is an n by m diagonal matrix of the form

$$S = \begin{bmatrix} \sigma_1 & 0 & 0 & 0 & 0 & 0 \\ 0 & \sigma_2 & 0 & 0 & 0 & 0 \\ 0 & 0 & \ddots & 0 & 0 & 0 \\ 0 & 0 & 0 & \sigma_r & 0 & 0 \\ 0 & 0 & 0 & 0 & 0 & 0 \\ 0 & 0 & 0 & 0 & 0 & 0 \end{bmatrix}$$

where $\sigma_1, \ldots, \sigma_r$ are positive numbers on the main diagonal with $\sigma_i \geq \sigma_{i+1}$. Constants σ_j are called the singular values with the number of nonzero values being equal to the rank r.

To understand the structure of this decomposition, let us study the case where $n \geq m$. Direct multiplication gives

$$A'AV = V \, \mathbf{diag}([\sigma_1^2, \ldots, \sigma_r^2, \mathbf{zeros}(1, m - r)])$$

and

$$AA'U = U \, \mathbf{diag}([\sigma_1^2, \ldots, \sigma_r^2, \mathbf{zeros}(1, n - r)])$$

Consequently, the singular values are square roots of the eigenvalues of the symmetric matrix $A'A$. Matrix V contains the orthonormalized eigenvectors arranged so that $\sigma_i \geq \sigma_{i+1}$. Although the eigenvalues of $A'A$ are obviously real, it may appear that this matrix could have some negative eigenvalues leading to pure imaginary singular values. However, this cannot happen because $A'AY = \lambda Y$ implies $\lambda = (AY)'(AY)/(Y'Y)$, which clearly is nonnegative. Once the eigenvectors and eigenvalues of $A'A$ are computed, columns of matrix U can be found as orthonormalized solutions of

$$[A'A - \sigma_j I]U_j = 0 \qquad \sigma_j = 0 \qquad j > r$$

The arguments just presented show that performing singular value decomposition involves solving a symmetric eigenvalue problem. However, SVD requires additional computation beyond solving a symmetric eigenvalue problem. It can be very time consuming for large matrices. The SVD has various uses, such as solving the normal equations. Suppose an n by m matrix A has $n > m$ and $r = m$. Substituting the SVD into

$$A'AX = A'B$$

gives

$$V \; \mathbf{diag}(\sigma_1^2, \ldots, \sigma_m^2)V'X = VS'U'B$$

Consequently, the solution of the normal equations is

$$X = V \; \mathbf{diag}(\sigma_1^{-1}, \ldots, \sigma_m^{-1})S'U'B$$

Another important application of the SVD concerns generation of orthonormal bases for the column space and the row space. The column space has dimension r and the null space has dimension $m - r$. Consider a consistent system

$$AX = B = U(SV'X)$$

Denote $SV'X$ as Y and observe that $y_j = 0$ for $j > r$ since $\sigma_j = 0$. Because B can be any vector in the column space, it follows that the first r columns of U, which are also orthonormal, are a basis for the column space. Furthermore, the decomposition can be written as

$$AV = US$$

This implies

$$AV_j = U_j\sigma_j = 0 \qquad j > r$$

which shows that the final $m - r$ columns of V form an orthonormal basis for the null space. The reader can verify that bases for the row space and left null space follow analogously by considering $A' = VS'U'$, which simply interchanges the roles of U and V.

MATLAB provides numerous other useful matrix decompositions such as LU, QR, and Cholesky. Some of these are employed in other sections of this book. The reader will find it instructive to read the built-in help information for MATLAB functions which describe these decomposition methods. For instance, the command **help** \ gives extensive documentation on the operation for matrix inversion.

3.7 Program Comparing FLOP Counts for Various Matrix Operations

This chapter is concluded with a program to measure the number of floating point operations, FLOPs, needed to perform various algebraic operations on real square matrices. These operations are approximately proportional to n^3 for sufficiently high matrix order. The program output for orders $n = 100$ and $n = 200$ is included in the next section. Notice that the SVD takes more than 23 times as much computation as Gauss reduction, and solving $AX = \lambda BX$ by the ZQ method takes 16 times as much computation as that required to solve $B^{-1}AX = \lambda X$.

MATLAB EXAMPLE

3.7.0.1   Output from Example

```
FLOP COUNT FOR VARIOUS MATRIX OPERATIONS


For n = 100


Gauss reduction                            =>    0.732 n^3
Cholesky decomposition                     =>    0.338 n^3
matrix multiplication                      =>    2.000 n^3
matrix inversion                           =>    2.040 n^3
solve A*X=(lambda)*X for symmetric A       =>    9.646 n^3
solve A*X=(lambda)*X for general A         =>   26.536 n^3
solve inv(B)*A*X=(lambda)*X                =>   24.738 n^3
solve A*X=(lambda)*B*X by QZ method        => 155.643 n^3
solve A*X=(lambda)*B*X for A,B symmetric=>    0.359 n^3
singular value decomposition  A=U*S*V'     =>   17.195 n^3


For n = 200


Gauss reduction                            =>    0.699 n^3
Cholesky decomposition                     =>    0.336 n^3
matrix multiplication                      =>    2.000 n^3
matrix inversion                           =>    2.020 n^3
solve A*X=(lambda)*X for symmetric A       =>    9.265 n^3
solve A*X=(lambda)*X for general A         =>   25.842 n^3
solve inv(B)*A*X=(lambda)*X                =>   23.475 n^3
solve A*X=(lambda)*B*X by QZ method        => 147.260 n^3
solve A*X=(lambda)*B*X for A,B symmetric=>    0.346 n^3
singular value decomposition  A=U*S*V'     =>   16.482 n^3
```

3.7.0.2 Script File flopex

```
 1: %   Example:  flopex
 2: %   ~~~~~~~~~~~~~~~~~
 3: %   This program tests the number of floating
 4: %   point operations required to perform several
 5: %   familiar matrix calculations. The operations
 6: %   tested are:
 7: %
 8: %      - solve A*X=B by Gauss reduction
 9: %      - solve A*X=B by Cholesky decomposition
10: %      - multiply square matrices
11: %      - invert a matrix
12: %      - eigenvectors and eigenvalues for a
13: %        symmetric matrix
14: %      - eigenvalues and eigenvectors of a real
15: %        matrix
16: %      - eigenvalues and eigenvectors of
17: %        A*X = Lambda*B*X  for A symmetric and B
18: %        symmetric positive definite
19: %      - eigenvalues of A*X=Lambda*B by the
20: %        QZ method
21: %      - singular value decomposition A=U*S*V'
22: %
23: %   User m functions required:
24: %        floptest, trisub
25: %-------------------------------------------------
26:
27: for j=100:50:200
28:   disp(' ');
29:   disp('Calculating: can take a while');
30:   floptest(j); clear;
31: end
```

3.7.0.3 Function floptest

```
 1: function [fcount,flist]=floptest(n)
 2: %
 3: % [fcount,flist]=floptest(n)
 4: % ~~~~~~~~~~~~~~~~~~~~~~~~~~~
 5: % This function determines the flop counts
 6: % needed to perform various matrix operations
```

```
 7: % which are proportional to n^3 for
 8: % sufficiently large n.
 9: %
10: %[fgus;fchol;fmlt;finv;fseig;feig;fsgeig;fgeig]
11: % are the flop count multipliers for various
12: % matrix operations
13: %
14: % User m functions called: none
15: %-----------------------------------------------
16:
17: a=rand(n,n)-.5; b=rand(n,n)-.5; c=a*ones(n,1);
18: as=a+a'; bs=b+b';
19: ev=sort(eig(bs));
20: bs=bs+(1+abs(ev(1)))*eye(n,n);
21: n3=n^3;
22:
23: % Solve A*X=C by Gauss reduction
24: fgus=flops; x=a\c; fgus=(flops-fgus)/n3; x=[];
25:
26: % Solve BS*X=C by Cholesky decomposition
27: fchol=flops; lo=chol(bs)';
28: x=trisub(lo,c); x=trisub(lo',x,1);
29: fchol=(flops-fchol)/n3; x=[];
30:
31: % Multiply two square matrices
32: fmlt=flops; d=a*b;
33: fmlt=(flops-fmlt)/n3; d=[];
34:
35: % Perform a matrix inversion
36: finv=flops; d=inv(a);
37: finv=(flops-finv)/n3; d=[];
38:
39: % Eigenvalues and eigenvectors for a real
40: % symmetric matrix
41: fseig=flops; [vecs,vals]=eig(bs);
42: fseig=(flops-fseig)/n3;
43: vecs=[]; vals=[];
44:
45: % Eigenvalues and eigenvectors for a general
46: % real matrix
47: feig=flops; [vecs,vals]=eig(a);
48: feig=(flops-feig)/n3;
49: vecs=[]; vals=[];
50:
51: % Eigenvalues and eigenvectors for
```

```
52: % A*X=lambda*B*X using inv(B)*A
53:
54: fgeig1=flops; [vecs,vals]=eig(b\a);
55: fgeig1=(flops-fgeig1)/n3;
56: vecs=[]; vals=[];
57:
58: % Eigenvalues and eigenvectors for
59: % A*X=lambda*B*X using the more
60: % conservative QZ algorithm
61:
62: fgeig2=flops; [vecs,vals]=eig(a,b);
63: fgeig2=(flops-fgeig2)/n3;
64: vecs=[]; vals=[];
65:
66: % Eigenvalues and eigenvectors for
67: % A*X=lambda*B*X where A is symmetric
68: % and B is symmetric positive definite
69:
70: fsgeig=flops; lo=chol(bs)'; aas=trisub(lo,as);
71: aas=trisub(lo,aas')'; aas=(aas+aas')/2;
72: [vecs,vals]=eig(aas); vecs=trisub(lo',vecs,1);
73: fsgeig=(flops-fsgeig)/n3; vecs=[]; vals=[];
74:
75: % Singular value decomposition of
76: % a square matrix
77:
78: fsvd=flops; [bu,bs,bv]=svd(b);
79: fsvd=(flops-fsvd)/n3;
80: s0 =' => %7.3f n^3\n';
81: s1 =['Gauss reduction                ', ...
82:      '               ',s0];
83: s2 =['Cholesky decomposition         ', ...
84:      '               ',s0];
85: s3 =['matrix multiplication          ', ...
86:      '             ',s0];
87: s4 =['matrix inversion               ', ...
88:      '              ',s0];
89: s5 =['solve A*X=(lambda)*X for symmetric', ...
90:      ' A       ',s0];
91: s6 =['solve A*X=(lambda)*X for general A', ...
92:      '             ',s0];
93: s7 =['solve inv(B)*A*X=(lambda)*X   ', ...
94:      '             ',s0];
95: s8 =['solve A*X=(lambda)*B*X by QZ method', ...
96:      '          ',s0];
```

```
97:  s9 =['solve A*X=(lambda)*B*X for A,B', ...
98:       ' symmetric ',s0];
99:  s10=['singular value decomposition   ', ...
100:      'A=U*S*V''     ',s0];
101:
102: %          1      2       3       4       5
103: fcount=[fgus;  fchol;  fmlt;   finv;   fseig;...
104:         feig;  fgeig1; fgeig2; fsgeig; fsvd];
105: %          6      7       8       9       10
106:
107: disp(' '),
108: fprintf('FLOP COUNT FOR VARIOUS ')
109: fprintf('MATRIX OPERATIONS\n\n')
110: disp(['For n = ',int2str(n)]), disp(' ')
111:
112: fprintf(s1,fgus),    fprintf(s2,fchol)
113: fprintf(s3,fmlt),    fprintf(s4,finv)
114: fprintf(s5,fseig),   fprintf(s6,feig)
115: fprintf(s7,fgeig1),  fprintf(s8,fgeig2)
116: fprintf(s9,fsgeig),  fprintf(s10,fsvd)
```

3.7.0.4   Function trisub

```
1:  function x=trisub(lowr,b,ifupper)
2:  %
3:  % x=trisub(lowr,b,ifupper)
4:  % ~~~~~~~~~~~~~~~~~~~~~~~~~
5:  % Solve LOWR*X = B, where LOWR is lower
6:  % triangular.  When ifupper is present, then
7:  % LOWR is assumed to be an upper triangular
8:  % matrix. Note that the right side matrix B can
9:  % have more than one column.
10: %
11: % User m functions called:  none
12: %-----------------------------------------------
13:
14: [n,m]=size(lowr); [nb,mb]=size(b);
15: x=zeros(n,mb);
16: if nargin==3
17:    nn=n:-1:1; lowr=lowr(nn,nn); b=b(nn,:);
18: end
19: x(1,:)=b(1,:)/lowr(1,1);
```

4

Methods for Interpolation and Numerical Differentiation

4.1 Concepts of Interpolation

Interpolation is a process whereby a function is approximated using data known at a discrete set of points. Typically we have points (x_i, y_i), $1 \leq i \leq n$, arranged such that $x_{i+1} > x_i$. These points are to be connected by a continuous interpolation function influenced by smoothness requirements such as:

a) the function should not deviate greatly from the data at points lying between the data values.

b) the function should satisfy a differentiability condition such as continuity of first and second derivatives.

In the most simple form of interpolation, we connect successive points by straight lines. This method, known as piecewise linear interpolation, satisfies condition (a) but has the disadvantage of producing piecewise constant slope values which yield slope discontinuities. An obvious cure for slope discontinuity is to use a curve such as a polynomial of degree $n - 1$ (through n points) to produce an interpolation function having all derivatives continuous. However, it was seen in Section 2.3 that a polynomial passing exactly through the data points may be highly irregular at intermediate values. Frequently, using polynomial interpolations higher than order five or six produces disappointing results. An excellent alternative to allowing either slope discontinuities or demanding slope continuity of all orders is to use cubic spline interpolation. This method connects successive points by cubic curves joined such that function continuity as well as continuity of the first two function derivatives is achieved.

MATLAB's intrinsic function **table1** performs piecewise linear interpolation and intrinsic function **spline** performs piecewise cubic interpolation with derivative continuity through order two. Another related function **linspace** is convenient for generating a set of equally spaced values between specified limits. For example, spline interpolating through 21 points on the sine curve and evaluating the spline at 101 points is accomplished by the statements

```
xd = linspace(0,2*pi,21);
x = linspace(0,2*pi,101);
y = spline(xd,sin(xd),x);
```

The piecewise linear interpolation function **table1** is restricted to handling points within the data range and does not allow jump discontinuities as is necessary to describe a function such as a sawtooth wave. Consequently, we have provided the following function **lintrp** which allows jump discontinuities characterized by a condition such as $x_{j+1} = x_j$, $y_{j+1} \neq y_j$. Furthermore, function values outside the original data range are evaluated using the first or last y value, whichever is closest. Function **lintrp** is employed subsequently in Chapter 6.

An auxilary library of MATLAB functions called "The Spline Toolbox" [84], is available from The MathWorks, Inc. This library includes the spline function as well as various other capabilities. However, the standard spline function intrinsic to MATLAB does not include spline differentiation and integration. To provide this capability we have developed two cubic spline functions **spc** and **spltrp** which are discussed in detail in this chapter and are used extensively in geometric property calculations presented in Chapter 5.

Before proceeding with the mathematical formulation for cubic splines, let us mention briefly the problem of passing a polynomial of degree $n - 1$ through n data points. In principle, polynomial interpolation can be readily accomplished with existing MATLAB functions with the statement

```
yinterp=polyval(...
        polyfit(xdata,ydata,length(xdata)-1),xinterp);
```

However, the resulting computations sometime lead to poorly conditioned simultaneous equations. A viable alternative is to use Newton's algorithm employing divided difference formulas developed in Conte and de Boor [17]. The following functions **polyterp**, **dvdcof**, and **dvdtrp** provide polynomial interpolation analogous to those given by **spline**. We will not discuss this method further because interpolation using high order polynomials usually yields less satisfactory results than is obtained with splines.

```
function y=lintrp(x,xd,yd)
%
% y=lintrp(x,xd,yd)
% ~~~~~~~~~~~~~~~~~~
% This function performs piecewise linear
% interpolation through data defined by vectors
% xd,yd. The components of xd are presumed to
% be in nondecreasing order. Any point where
% xd(i)==xd(i+1) generates a jump
% discontinuity. For points outside the data
% range, the interpolation is based on the
% lines through the outermost pairs of points
% at each end.
%
% x     - vector of values for which piecewise
%          linear interpolation is required
% xd,yd - vectors of data values through which
%          interpolation is performed
% y     - interpolated function values for
%          argument x
%
% User m functions required: none
%-------------------------------------------------
```

```
x=x(:); xd=xd(:); yd=yd(:);
y=zeros(length(x),1);
xmin=min(x); xmax=max(x); nd=length(xd);

if xmax > xd(nd)
  yd=[yd; yd(nd)+(yd(nd)-yd(nd-1))/ ...
     (xd(nd)-xd(nd-1))*2*(xmax-xd(nd))];
  xd=[xd;2*xmax-xd(nd)]; nd=nd+1;
end

if xmin < xd(1)
  yd=[yd(1)+(yd(2)-yd(1))/(xd(2)-xd(1))* ...
     (xmin-xd(1));yd];
  xd=[xmin;xd]; nd=nd+1;
end

for i=1:nd-1
  xlft=xd(i); ylft=yd(i);
  xrht=xd(i+1); yrht=yd(i+1); dx=xrht-xlft;
  if dx~=0, s=(yrht-ylft)/dx;
    y=y+(x>=xlft).*(x<xrht).* ...
      (ylft+s*(x-xlft));
  end
end

k=find(x==xd(nd));
if length(k)>0, y(k)=yd(nd); end
```

4.1.1   EXAMPLE: NEWTON POLYNOMIAL INTERPOLATION

4.1.1.1   Function polyterp

```
1: function yterp=polyterp(xdat,ydat,xterp)
2: %
3: % yterp=polyterp(xdat,ydat,xterp)
4: % ~~~~~~~~~~~~~~~~~~~~~~~~~~~~~~~~~
5: % Interpolate through n=length(xdat) data
6: % points using a polynomial of degree n-1.
7: %
8: % User m functions called:  dvdtrp, dvdcof
9: %-----------------------------------------------
10:
11: yterp=dvdtrp(xdat,dvdcof(xdat,ydat),xterp);
```

4.1.1.2   Function dvdcof

```
1: function c=dvdcof(xdat,ydat)
2: %
3: % c=dvdcof(xdat,ydat)
4: % ~~~~~~~~~~~~~~~~~~~
5: % This function uses divided differences to
6: % compute coefficients c needed to perform
7: % polynomial interpolation by the Newton form
8: % of the interpolating polynomial. The data
9: % values used are xdat(i),ydat(x), i=1:n.
10: %
11: % Reference:
12: %    "Elementary Numerical Analysis"
13: %    by S. Conte and C. de Boor
14: %
15: % User m functions called:  none.
16: %-----------------------------------------------
17:
18: n=length(xdat);
19: for k=1:(n-1)
20:    for i=1:(n-k)
21:       ydat(i)=...
22:       (ydat(i+1)-ydat(i))./(xdat(i+k)-xdat(i));
23:    end
```

```
24: end
25: c=ydat;
```

4.1.1.3 Function dvdtrp

```
 1: function yterp=dvdtrp(xdat,c,xterp)
 2: %
 3: % yterp=dvdtrp(xdat,c,xterp)
 4: % ~~~~~~~~~~~~~~~~~~~~~~~~~~
 5: % This function performs polynomial
 6: % interpolation using the Newton form of
 7: % the interpolating polynomial. The
 8: % coefficients c are first computed by a
 9: % call to function dvdcof. The
10: % points xdat(i) are abscissas of the data
11: % values.  The vector yterp is returned as the
12: % interpolated function for argument values
13: % defined by vector xterp.
14: %
15: % Reference:
16: %    "Elementary Numerical Analysis"
17: %    by S. Conte and C. de Boor
18: %
19: % User m functions called:  none.
20: %------------------------------------------------
21:
22: xdat=xdat(:); c=c(:);
23: xterp=xterp(:);
24: n=length(xterp); m=length(xdat);
25: yterp=c(1)*ones(n,1);
26: for i=2:m
27:   yterp=c(i)+(xterp-xdat(i)).*yterp;
28: end
```

4.2   Interpolation, Differentiation, and Integration By Cubic Splines

Cubic spline interpolation is a versatile method commonly employed to pass a smooth curve through a sequence of data points. The technique connects each successive pair of points using a cubic polynomial. Boundary conditions are imposed to make $y(x)$, $y'(x)$, and $y''(x)$ continuous whenever contiguous intervals join. Consequently, the resulting piecewise cubic curve has continuous derivatives through order two. Therefore, $y(x)$, $y'(x)$, $y''(x)$, and $y'''(x)$ are, respectively, piecewise cubic, piecewise parabolic, piecewise linear, and piecewise constant. Once the polynomials for different intervals have been calculated, the function, the two derivatives and the integral can be evaluated. Two functions are developed below to perform spline interpolation. These functions extend the intrinsic spline functions of MATLAB to include differentiation and integration. In the first calculation phase, the values of $y''(x)$ needed to assure slope continuity are determined. This set of second derivatives is subsequently used as coefficients of the cubic polynomial for each interval. The coefficients completely define the interpolation curve. They may also be employed to evaluate the spline at any number of points. Details of the procedure necessary to perform these operations are described next. Readers who want further detail on spline theory will find the books by de Boor [23] and Ahlberg and Nilson [2] to be comprehensive references.

First, we discuss how to compute a cubic polynomial where $y(x)$ and $y''(x)$ are specified at $x = 0$ and $x = h_1$ (see Figure 4.1). The values at the left and right ends are

$$y(0) = y_1 \qquad y''(0) = T_1 \qquad \text{left end}$$

$$y(h_1) = y_2 \qquad y''(h_1) = T_2 \qquad \text{right end}$$

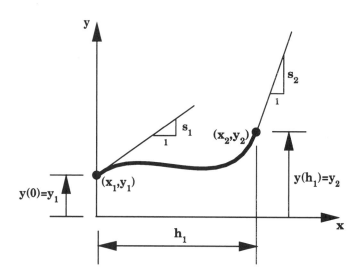

FIGURE 4.1. Cubic Segment

A Taylor series expansion implies that

$$y(x) = y_1 + s_1 x + \frac{1}{2}T_1 x^2 + \frac{1}{6}V_1 x^3$$

where

$$s_1 = y'(0) \qquad V_1 = y'''(0)$$

Since $y''(x)$ is a linear function with

$$y''(0) = T_1 \qquad y''(h_1) = T_2$$

then

$$y''(x) = T_1 + (T_2 - T_1)\frac{x}{h_1}$$

and

$$y'''(x) = (T_2 - T_1)\frac{1}{h_1}$$

Consequently,

$$y(x) = y_1 + s_1 x + \frac{1}{2}T_1 x^2 + \left(\frac{T_2 - T_1}{6h_1}\right) x^3$$

The imposed function value at $x = h_1$ requires

$$y_2 = y_1 + s_1 h_1 + \frac{(2T_1 + T_2)h_1}{6}$$

Therefore, the slope at $x = 0$ satisfies

$$y'(0) = s_1 = \frac{y_2 - y_1}{h_1} - \frac{(2T_1 + T_2)h_1}{6}$$

and, similarly, the slope at the right end is given by

$$y'(h_1) = s_2 = \frac{y_2 - y_1}{h_1} + \frac{(T_1 + 2T_2)h_1}{6}$$

The conditions necessary when contiguous cubic curves have the same slope and function values at the common interface point may now be developed. Suppose the left segment has length h_1 and satisfies $y = y_1$, $y'' = T_1$ at the left end, and $y = y_2$, $y'' = T_2$ at the right end. Similarly, the right segment of length h_2 satisfies $y = y_2$, $y'' = T_2$ at the left end, and $y = y_3$, $y'' = T_3$ at the right end. The equations developed above show that making the slope at the right end of the left interval match the slope at the left end of the right interval requires

$$\frac{y_2 - y_1}{h_1} + \frac{(T_1 + 2T_2)h_1}{6} = \frac{y_3 - y_2}{h_2} - \frac{(2T_2 + T_3)h_2}{6}$$

Consequently, the second derivatives T_1, T_2, and T_3 are connected by

$$h_1 T_1 + 2(h_1 + h_2)T_2 + h_2 T_3 = 6\left[\frac{y_2 - y_1}{h_1} - \frac{y_3 - y_2}{h_2}\right]$$

The last relation can be generalized to handle a problem involving several intervals. If n data points are given as

$$x = x_j \qquad y = y_j \qquad x_j > x_{j-1} \qquad 1 \le j \le n$$

then $T_1, \ldots, T_n$ must satisfy

$$h_{j-1}T_{j-1} + 2(h_{j-1} + h_j)T_j + h_j T_{j+1} =$$

$$6\left[\frac{y_j - y_{j-1}}{h_{j-1}} - \frac{y_{j+1} - y_j}{h_j}\right] \qquad 2 \le j \le n - 1$$

These $n - 2$ slope continuity conditions for interior points must be supplemented by end conditions at the left-most point x_1 and the right-most point x_n. The three types of conditions typically imposed are

a) slope, or

b) second derivative, or

c) continuity of $y'''(x)$ at the interior points nearest to the ends of the interval.

Boundary condition (c), pertaining to third order derivative continuity, has the effect of transferring information describing the function behavior at the interior points to the outside points. Since the value of $y'''(x)$ is constant for each subinterval, the requirement for $y'''(x_2)$ to be continuous is

$$\frac{T_2 - T_1}{h_1} = \frac{T_3 - T_2}{h_2}$$

which reduces to

$$h_2 T_1 - (h_1 + h_2)T_2 + h_3 T_3 = 0$$

The formulas developed above show that the types of boundary conditions applicable at the left end require

- for given slope:

$$2T_1 + T_2 = 6\left[\frac{y_2 - y_1}{h_1^2} - \frac{y'(x_1)}{h_1}\right]$$

- for given second derivative:

$$T_1 = y''(x_1)$$

- for continuous $y'''(x_2)$:

$$h_2 T_1 - (h_1 + h_2)T_2 + h_1 T_3 = 0$$

Similar boundary conditions imposed at $x = x_n$ yield

- for given slope:

$$T_{n-1} + 2T_n = 6\left[-\frac{y_n - y_{n-1}}{h_{n-1}^2} + \frac{y'(x_n)}{h_{n-1}}\right]$$

- for given second derivative:

$$T_n = y''(x_n)$$

- for continuous $y'''(x_{n-1})$:

$$h_{n-1}T_{n-2} - (h_{n-2} + h_{n-1})T_{n-1} + h_{n-2}T_n = 0$$

This set of relationships yields a sparse system of simultaneous equations solvable for $T_1, \ldots, T_n$. For example, when end conditions specifying slope or second derivative are used, the system has the tridiagonal form

$$a_j x_{j-1} + b_j x_j + c_j x_{j+1} = d_j$$

This system can be solved by Gauss reduction, or

```
for k = 1:n-1
  r = a(k+1) / b(k);
  b(k+1) = b(k+1) - r * c(k);
  d(k+1) = d(k+1) - r * d(k);
end
x(n) = d(n) / b(n);
for k = n-1:-1:1
  x(k) = ( d(k) - c(k) * x(k+1) ) / b(k);
end
```

The above code, which must be processed by the MATLAB interpreter, can be replaced with the more efficient intrinsic equation solver which has been compiled and therefore runs much faster than interpreted code.

Two MATLAB functions **spc** and **spltrp** are described below. Function **spc** accepts input data coordinates (x, y) and end conditions specified by vectors i and v. Constant $v(1)$ contains a value for the left end slope or deflection when $i(1)$ equals 1 or 2. The value of $v(1)$ is ignored when $i(1)$ equals 3. Parameters $i(2)$ and $v(2)$ impose right end conditions analogous to what $i(1)$ and $v(1)$ accomplish at the left end. One other argument is the vector $icrnr$ specifying indices of any interior points where the slope continuity condition is replaced by a condition requiring $y''(x)$ to be zero.

This type of condition is comparable to placing a hinge in a deflected beam to create what is referred to here as a corner point (a point where the curve slope experiences a finite jump). No slope discontinuities occur if $icrnr$ has zero length or is omitted from the argument list. It is worth noting that piecewise linear interpolation is implied when a spline curve has $y''(x) = 0$ at both ends and all interior points are corner points. The final output from **spc** is matrix $splmat$ which contains all the data needed by function **spltrp** to interpolate, differentiate, or integrate the spline from x_1 to an arbitrary upper limit.

Function **spltrp** accepts input consisting of an array, mat, and a vector of arguments, x, where the spline interpolation is desired. Parameter $ideriv$ has values of 0, 1, 2, or 3. The value of $ideriv$ selects whether values of $y(x), y'(x), y''(x)$, or the integral from x_1 to x are requested. Another special characteristic of function **spltrp** is that function values outside of the original data range are based on the corresponding end tangents at x_1 and x_n. This choice seems preferable because evaluating the polynomials for the outer segments often produces very large function values at points outside the original data range.

Two examples illustrating spline interpolation are presented next. In the first example, a series of equally spaced points between 0 and 2π is used to approximate $y(x) = \sin(x)$. This function satisfies

$$y'(x) = \cos(x)$$
$$y''(x) = -\sin(x)$$
$$\int_0^x \sin(x)dx = 1 - \cos(x)$$

The approximations for the function, derivatives, and the integral shown in Figure 4.2 are quite satisfactory. The script file which produces this figure is listed as **sinetrp**.

In the second example, interpolation of a two-dimensional space curve where y cannot be interpolated as a single valued function of x is studied. In this case, a parameter t_j is employed which has a value equal to the index j for each (x_j, y_j) used. Then $x(t)$ and $y(t)$ are interpolated to produce a smooth curve through the data. A function **spcurv2d** is provided to compute points on a plane curve. Another script file named **matlbdat** produces data points spelling the word MATLAB when connected. Results shown in

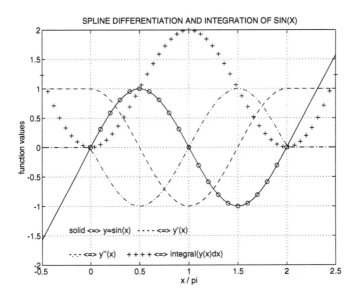

FIGURE 4.2. Spline Differentiation and Integration of $\sin(x)$

Figure 4.3 illustrate the capability which splines provide to describe a complicated curved shape. The use of corner points is critical in this example to make the sharp turns needed to describe letters such as the "t".

It should be mentioned in conclusion that functions **spc** and **spltrp** are more general than the intrinsic spline function in MATLAB. That function does not provide for corner points, control of end conditions, or allow integration and differentiation.

112

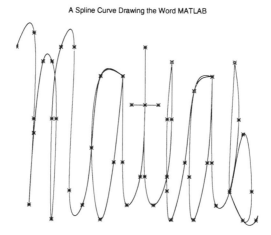

A Spline Curve Drawing the Word MATLAB

FIGURE 4.3. A Spline Curve Drawing the Word MATLAB

4.2.1   EXAMPLE: SPLINE INTERPOLATION APPLIED TO $\sin(x)$

4.2.1.1   Script File sinetrp

```
 1: % Example: sinetrp
 2: % ~~~~~~~~~~~~~~~~~
 3: % This example illustrates cubic spline
 4: % approximation of sin(x), its first two
 5: % derivatives, and its integral.
 6: %
 7: % User m functions required:
 8: %    spc, spltrp, sinetrp
 9: %-----------------------------------------------
10:
11: % Obtain data points on the spline curve
12: x=linspace(0,2*pi,21); y=sin(x);
13: xx=linspace(-pi/2,2.5*pi,51);
14: % Specify y' at first point and
15: % y'' at last point
16: i=[1,2]; v=[1,0];
17: % Get interpolation coefficients
18: splmat=spc(x,y,i,v);
19:
20: % Evaluate function values at a dense
21: % set of points
22: xx=linspace(-pi/2,2.5*pi,51);
23: z=xx/pi;
24: yy=spltrp(xx,splmat,0);
25: yyp=spltrp(xx,splmat,1);
26: yypp=spltrp(xx,splmat,2);
27: yyint=spltrp(xx,splmat,3);
28:
29: % Plot results
30: plot(x/pi,y,'ow',z,yy,'-w',z,yyp,'--w',...
31:      z,yypp,'-.w',z,yyint,'+w'), grid
32: title(['SPLINE DIFFERENTIATION AND ', ...
33:        'INTEGRATION OF SIN(X)'])
34: xlabel('x / pi'), ylabel('function values')
35: text(-.25,-1.40,...
36: ' solid <=> y=sin(x)   - - - - <=> y''(x)')
37: str=[' -.-.- <=> y''''(x)      + + + + ', ...
38:        '<=> integral(y(x)dx)'];
39: text(-.25,-1.80,str)
```

4.2.2 EXAMPLE: PLOTTING OF GENERAL PLANE CURVES

4.2.2.1 Script File matlbdat

```
 1: % Example: matlbdat
 2: % ~~~~~~~~~~~~~~~~~~
 3: % This example illustrates the use of splines
 4: % to draw the word MATLAB.
 5: %
 6: % User m functions required:
 7: %      spcurv2d, spc, spltrp
 8: %-----------------------------------------------
 9:
10: x=[13 17 17 16 17 19 21 22 21 21 23 26
11:    25 28 30 32 37 32 30 32 35 37 37 38
12:    41 42 42 42 45 39 42 42 44 47 48 48
13:    47 47 48 51 53 57 53 52 53 56 57 57
14:    58 61 63 62 61 64 66 64 61 64 67 67];
15: y=[63 64 58 52 57 62 62 58 51 58 63 63
16:    53 52 56 61 61 61 56 51 55 61 55 52
17:    54 59 63 59 59 59 59 54 52 54 58 62
18:    58 53 51 55 60 61 60 54 51 55 61 55
19:    52 53 58 62 53 57 53 51 53 51 51 51];
20: x=x'; x=x(:); y=y'; y=y(:);
21: ncrnr=[17 22 26 27 28 29 30 31 36 42 47 52];
22: clf; [xs,ys]=spcurv2d(x,y,10,ncrnr);
23: plot(xs,ys,'w',x,y,'*w'), axis off
24: title('A Spline Curve Drawing the Word MATLAB')
```

4.3   Numerical Differentiation Using Finite Differences

Problems involving differential equations are often solved approximately by using difference formulas which approximate the derivatives in terms of function values at adjacent points. Deriving difference formulas by hand can be tedious, particularly when unequal point spacing is used. For this reason, we develop a numerical procedure allowing construction of formulas of arbitrary order and arbitrary truncation error. Of course, as the desired order of derivative and the order of truncation error increases, a larger number of points is needed to interpolate the derivative. We will show below that approximating a derivative of order k with a truncation error of order h^m generally requires $(k + m)$ points unless symmetric central differences are used.

Consider the Taylor series expansion

$$F(x + \alpha h) = \sum_{k=0}^{\infty} \frac{F^{(k)}(x)}{k!} (\alpha h)^k$$

where $F^{(k)}(x)$ means the k'th derivative of $F(x)$. This relation expresses values of F as linear combinations of the function derivatives at x. Conversely, the derivative values can be cast in terms of function values by solving a system of simultaneous equations. Let us take a series of points defined by

$$x_i = x + h\alpha_i \qquad 1 \le i \le n$$

where h is a fixed step-size and α_i are arbitrary parameters. Separating some leading terms in the series expansion gives

$$F(x_i) = \sum_{k=0}^{n-1} \frac{\alpha_i^k}{k!} \left[h^k F^{(k)}(x) \right] + \frac{\alpha_i^n}{n!} \left[h^n F^{(n)}(x) \right] +$$

$$\frac{\alpha_i^{n+1}}{(n+1)!} \left[h^{(n+1)} F^{(n+1)}(x) \right] + O(h^{n+2}) \qquad 1 \le i \le n$$

It is helpful to use the following notation:

α^k – a column vector with component i being equal to α_i^k

f – a column vector with component i being $F(x_i)$

fp – a column vector with component i being $h^i F^{(i)}(x)$

A – $[\alpha^0, \alpha^1, \ldots, \alpha^{n-1}]$, a square matrix with columns which are powers of α

Then the Taylor series expressed in matrix form is

$$f = A * fp + \frac{h^n F^{(n)}(x)}{n!} \alpha^n +$$

$$\frac{h^{n+1} F^{(n+1)}(x)}{(n+1)!} \alpha^{n+1} + O(h^{n+2})$$

Solving this system for the derivative matrix fp yields

$$fp = A^{-1} f - \frac{h^n F^{(n)}(x)}{n!} A^{-1} \alpha^n -$$

$$\frac{h^{n+1} F^{(n+1)}(x)}{(n+1)!} A^{-1} \alpha^{n+1} + O(h^{n+2})$$

In the last equation we have retained the first two remainder terms in explicit form to allow the magnitudes of these terms to be examined. Row $k + 1$ of the previous equation implies

$$F^{(k)}(x) = h^{-k}(A^{-1}f)_{k+1} - \frac{h^{n-k}}{n!} F^{(n)}(x)(A^{-1}\alpha^n)_{k+1} -$$

$$\frac{h^{n-k+1}}{(n+1)!} F^{(n+1)}(x)(A^{-1}\alpha^{n+1})_{k+1} + O(h^{n-k+1})$$

Consequently, the rows of A^{-1} provide coefficients in formulas to interpolate derivatives. For a particular number of interpolation points, say N, the highest derivative approximated will be $F^{(N-1)}(x)$ and the truncation error will normally be of order k^1. Conversely, if we need to compute a derivative formula of order k with the truncation error being m, then it is necessary to use a number of points such that $n - k = m$; therefore $n = m + k$. For the case where interpolation points are symmetrically placed around the point where derivatives are desired, one higher power of accuracy order is achieved than might be expected. We can show, for example, that

$$\frac{d^4 F(x)}{dx^4} = \frac{1}{h^4}[F(x - 2h) - 4F(x - h) + 6F(x) -$$

$$4F(x + h) + F(x + 2h)] + O(h^2)$$

because the truncation error term associated with h^1 is found to be zero. At the same time, we can show that a forward difference formula for $f'''(x)$ employing equidistant point spacing is

$$\frac{d^3 F(x)}{dx^3} = \frac{1}{h^3}[-2.5F(x) - 9F(x + h) + 12F(x + 2h) +$$

$$7F(x + 3h) - 1.5F(x + 4h)] + O(h^2)$$

Although the last two formulas contain arithmetically simple interpolation coefficients, due to equal point spacing, the method is certainly not restricted to equal spacing. The following program contains the function **derivtrp** which implements the ideas just developed. Since the program contains documentation which is output when it is executed, no additional example problem is included.

4.3.1 EXAMPLE: DERIVING GENERAL DIFFERENCE FORMULAS

4.3.1.1 Output from Example

```
>> finidif

FINITE DIFFERENCE FORMULAS OF ARBITRARY ORDER

select the derivative order and
the truncation error order
(input 0,0 to stop)
 ? 3,2
give a 5 component vector defining the base points
 ? -3,-2,-1,0,1

derivative  interpolation coefficients
    order  c(1)     c(2)     c(3)     c(4)     c(5)
        0  0        0        0        1.0000   0
   1.0000 -0.0833   0.5000  -1.5000   0.8333   0.2500
   2.0000 -0.0833   0.3333   0.5000  -1.6667   0.9167

/---------------- Editorial Note ----------------\
|                                                 |
|    The following line gives the coefficients    |
|    to evaluate the third derivative using       |
|    five function values.                        |
|                                                 |
|   3.0000  0.5000 -3.0000  6.0000 -5.0000  1.5000|
|           ------  ------   ------  ------  ------|
\-------------------------------------------------/

   4.0000  1.0000 -4.0000   6.0000 -4.0000   1.0000

derivative    power of    error    power of    error
    order       h^k        coef.      h^k       coef.
        0      5.0000         0      6.0000         0
   1.0000      4.0000    0.0500      5.0000   -0.0417
   2.0000      3.0000    0.0833      4.0000   -0.0528
```

```
/---------------- Editorial Note -----------------\
|                                                  |
|      The truncation error coefficients of        |
|        order h^2 and h^3 are shown below         |
|                                                  |
|   3.0000    2.0000    -0.2500    3.0000   0.2500|
|                        ------              ------|
\--------------------------------------------------/
```

```
    4.0000    1.0000    -1.0000    2.0000   0.6667
```

select the derivative order and
the truncation error order
(input 0,0 to stop)
 ? 4,1
give a 5 component vector defining the base points
 ? -2,-1,0,1,2

derivative   interpolation coefficients
     order   c(1)   c(2)   c(3)   c(4)   c(5)
         0   0   0   1.0000   0   0
    1.0000   0.0833 -0.6667   0   0.6667 -0.0833
    2.0000 -0.0833   1.3333 -2.5000   1.3333 -0.0833
    3.0000 -0.5000   1.0000   0 -1.0000   0.5000
    4.0000   1.0000 -4.0000   6.0000 -4.0000   1.0000

derivative   power of   error   power of   error
     order   h^k   coef.   h^k   coef.
         0   5.0000   0   6.0000   0
    1.0000   4.0000   -0.0333   5.0000   0
    2.0000   3.0000   0   4.0000   -0.0111
    3.0000   2.0000   0.2500   3.0000   0
    4.0000   1.0000   0   2.0000   0.1667
```

select the derivative order and
the truncation error order
(input 0,0 to stop)
 ? 0,0
```

4.3.1.2 Script file finidif

```
 1: % Example: finidif
 2: % ~~~~~~~~~~~~~~~~~
 3: % This program uses a truncated Taylor series
 4: % to compute finite difference formulas
 5: % approximating derivatives of arbitrary order
 6: % which are interpolated at an arbitrary set of
 7: % base points not necessarily evenly spaced.
 8: % The base point positions are expressed as
 9: % multiples of a stepsize variable h. For
10: % example, if function values at x1=x-h, x2=x,
11: % and x3=x+h are used to evaluate y''(x), then
12: % the base points would be defined by the
13: % vector [-1,0,1]. The corresponding
14: % approximation for y''(x) would be expressed
15: % as
16: %
17: %        ( c(1)*y1 + c(2)*y2 + c(3)*y3 )/h^2
18: %
19: % Furthermore, the truncation error
20: % coefficients for the next two terms in the
21: % Taylor series expansion are included in the
22: % table of printed output. Results are given
23: % in a tableau illustrated by the following
24: % examples pertaining to third and fourth
25: % derivative formulas.
26: %
27: % User m functions required:
28: %    derivtrp, readv, output
29: %-------------------------------------------------
30:
31: disp('')
32: fprintf('\nFINITE DIFFERENCE FORMULAS ')
33: fprintf('OF ARBITRARY ORDER\n\n')
34: disp('')
35: while 1
36:   disp('select the derivative order and')
37:   disp('the truncation error order')
38:   disp('(input 0,0 to stop)');
39:   t=readv(2); n=sum(t); in=int2str(n);
40:   if n==0; break; end
41:   disp(' '); disp(['give a ',in,...
42: ' component vector defining the base points'])
```

```
43:   basepts=readv(n);
44:   [cof,ct,cb,e1,e2]=derivtrp(basepts);
45:   output(cof,e1,e2)
46: end
```

4.3.1.3   Function derivtrp

```
1: function [c,ctop,cbot,erc1,erc2]=...
2:            derivtrp(baspts)
3: %
4: % [c,ctop,abot,erc1,erc2]=derivtrp(baspts)
5: % ~~~~~~~~~~~~~~~~~~~~~~~~~~~~~~~~~~~~~~~~~~
6: % This function computes coefficients to
7: % interpolate derivatives by finite
8: % differences.
9: %
10: % baspts - The matrix specifying the base
11: %          points for which function values
12: %          are employed.  For example, using
13: %          function values of f(x-2h),
14: %          f(x-h), f(x), f(x), f(x+2h) and
15: %          f(x+3h) would imply baspts=[-2:3].
16: % c      - Matrix in which row k+2 contains
17: %          coefficients which properly weight
18: %          the function values corresponding
19: %          to base points defined by vector
20: %          baspts. The first row of c contains
21: %          the components of alp to help
22: %          identify which coefficients go with
23: %          which function values.
24: % ctop,  - These two matrices contain integers
25: % cbot     such that c is closely approximated
26: %          by ctop./cbot. When the difference
27: %          coefficients turn out to be exactly
28: %          expressible as rational functions,
29: %          then matrices ctop and cbot usually
30: %          give the desired coefficients
31: %          exactly.
32: % erc1,  - Vectors characterizing the
33: %          truncation error terms
34: % erc2     associated with various derivatives.
35: %
36: % User m functions called:  none.
```

```
37: %-----------------------------------------------
38:
39: x=baspts(:); n=length(x); a=ones(n,n+2);
40: for k=2:n+2;
41:   a(:,k)=a(:,k-1).*x/(k-1);
42: end
43: c=inv(a(:,1:n));
44: erc1=c*a(:,n+1); erc2=c*a(:,n+2);
45: [ctop,cbot]=rat(c);
```

4.3.1.4   Function output

```
1: function output(a,ercof1,ercof2)
2: %
3: % output(a,ercof1,ercof2)
4: % ~~~~~~~~~~~~~~~~~~~~~~~~
5: % This function prints the results.
6: %
7: % User m functions called:  none.
8: %-----------------------------------------------
9:
10: n=max(size(a)); mat1=[(0:n-1)',a];
11: mat2=[(0:n-1)',(n:-1:1)',ercof1(:), ...
12:       (n+1:-1:2)',ercof2];
13: s=' 	      order';
14: for j=1:n;
15:   s=[s,'        c(',int2str(j),')'];
16: end
17: fprintf('\nderivative     interpolation ')
18: fprintf('coefficients\n')
19: disp(s), disp(mat1)
20: if n>6, pause, end
21: fprintf('\nderivative     power of  error')
22: fprintf('     power of   error    ')
23: fprintf('\n    order       h^k      coef.')
24: fprintf('        h^k      coef.  \n')
25: disp(mat2)
```

4.3.2   EXAMPLE: DERIVING ADAMS TYPE INTEGRATION FORMULAS

The same ideas employed to derive finite difference relations can be applied to develop Adams type integration formulas used to integrate differential equations. Integrating the differential equation $y'(x) = f(x, y(x))$ over a time step gives

$$y(x + h) = y(x) + h \sum_{k=0}^{\infty} \left[\frac{f^{(k)}(x, y(x))}{(k+1)!} \right] h^k$$

Expressing the derivatives $f^{(k)}$ in terms of f values in a manner similiar to what was done earlier leads to formulas of the form

$$y_{i+1} = y_i + h \sum_{j} c_{ij} f_j$$

involving values of the derivative function $f(x, y(x))$ at selected points. The resulting formulas are either explicit if y_{i+1} is not involved in the points where f is evaluated or implicit if $f(x_{i+1}, y_{i+1})$ is present on the right side. The following short program computes Adams integration formulas. Two examples involving explicit and implicit formulas are given. Output produced by the program is also provided. This example concludes the current chapter on interpolation and differentiation.

MATLAB EXAMPLE

4.3.2.1 Output from Example

*** Adams Integration Formulas for DE Solution ***

Adams-Bashforth order six explicit

Index	C-top	C-bottom
0	199	67
-1	-1326	241
-2	1324	191
-3	-968	191
-4	959	480
-5	-95	288

Adams-Moulton order six implicit

Index	C-top	C-bottom
1	95	288
0	439	443
-1	-133	240
-2	241	720
-3	-173	1440
-4	3	160

4.3.2.2   Script file adamsex

```
 1: % Example:  adamsex
 2: % ~~~~~~~~~~~~~~~~~~
 3: % This program illustrates use of function
 4: % adams2 for determining coefficients in the
 5: % explicit or implicit Adams type formulas
 6: % used for differential equation solution.
 7: % These integration formulas have the form
 8: %
 9: %      ynplus1 = yn + h*Sum(cof(j)*f(j))
10: %
11: % where f(j) denotes the value of f for
12: %
13: %      x(j) = xn + h*alpha(j)
14: %
15: % The weighting coefficient cof(j) equals
16: % the quotient c(j,1)/c(j,2).
17: %
18: % User m functions required:
19: %     adams2
20: %--------------------------------------------
21:
22: fprintf('\n*** Adams Integration Formulas')
23: fprintf(' for DE Solution ***\n')
24: % order6_explicit=[(0:-1:-5)',adams2(0:-1:-5)]
25: % order6_implicit=[(1:-1:-4)',adams2(1:-1:-4)]
26: fprintf(...
27:    '\n        Adams-Bashforth  order six explicit')
28: fprintf(...
29:    '\n            Index     C-top     C-bottom \n')
30: disp([(0:-1:-5)',adams2(0:-1:-5)])
31: fprintf(...
32:    '\n        Adams-Moulton  order six implicit')
33: fprintf(...
34:    '\n            Index     C-top     C-bottom \n')
35: disp([(1:-1:-4)',adams2(1:-1:-4)])
```

4.3.2.3   Function adams2

```
 1: function c = adams2(alpha)
 2: %
```

```
 3: % c = adams2(alpha)
 4: % ~~~~~~~~~~~~~~~~~
 5: % This function determines coefficients in the
 6: % Adams type formulas used to solve
 7: % y'(x)=f(x,y).  These integration formulas
 8: % have the form
 9: %
10: %      ynplus1 = yn + h*Sum(cof(j)*f(j))
11: %
12: % where f(j) denotes the value of f for
13: %
14: %      x(j) = xn + h*alpha(j)
15: %
16: % The weighting coefficient cof(j) equals
17: % the quotient c(j,1)/c(j,2). For example,
18: % adams2(0:-1:-4) determines the Adams-
19: % Bashforth formula of order 5. Similarly,
20: % adams2(1:-1:-3) determines the Adams-Moulton
21: % formula of order 5.
22: %
23: % alpha - vector of coefficients defining the
24: %         base points used for the integration
25: % c     - weighting coefficients for the
26: %         integration formula
27: %
28: % User m functions called:  none.
29: %-----------------------------------------------
30:
31: alpha=alpha(:)'; n=length(alpha);
32: a=(ones(n,1)*alpha).^((0:n-1)'*ones(1,n));
33: [coftop,cofbot]=rat(a\(1 ./(1:n)'));
34: c=[coftop,cofbot];
```

5

Gaussian Integration with Applications to Geometric Properties

5.1 Fundamental Concepts and Intrinsic Integration Tools Provided in MATLAB

Numerical integration methods approximate a definite integral by evaluating the integrand at several points and taking a weighted combination of those integrand values. The weight factors are usually obtained by interpolating the integrand at selected points and integrating the interpolating function exactly. For example, the Newton-Cotes formulas result from polynomial interpolation through equidistant base points. This chapter presents concepts of numerical integration adequate to handle many practical applications.

Let us assume that an integral over limits a to b is to be evaluated. We can write

$$\int_a^b f(x)dx = \sum_{i=1}^n W_i f(x_i) + E$$

where E represents the error due to replacement of the integral by a sum. This is called an n-point quadrature formula. The points x_i where the integrand is evaluated are base points and the constants W_i are weight factors. Most integration formulas depend on approximating the integrand by a polynomial. Consequently, they give exact results when the integrand is a polynomial of sufficiently low order. Different choices of x_i and W_i will be discussed below.

It is helpful to express an integral over general limits in terms of some fixed limits, say -1 to 1. This is accomplished by introducing a linear change of variables

$$x = \alpha + \beta t$$

Requiring that $x = a$ corresponds to $t = -1$ and $x = b$ corresponds to $t = 1$ gives $\alpha = (a+b)/2$ and $\beta = (b-a)/2$, so that one obtains the identity

$$\int_a^b f(x)dx = \frac{1}{2}(b-a)\int_{-1}^1 f\left[\frac{a+b}{2} + \frac{b-a}{2}t\right]dt = \int_{-1}^1 F(t)dt$$

where $F(t) = f[(a+b)/2 + (b-a)t/2](b-a)/2$. Thus, the dependence of the integral on the integration limits can be represented parametrically by modifying the integrand. Consequently, if an integration formula is known for limits -1 to 1, we can write

$$\int_a^b f(x)dx = \beta \sum_{i=1}^n W_i f(\alpha + \beta x_i) + E$$

The idea of shifting integration limits can be exploited further by dividing the interval a to b into several parts and using the same numerical integration formula to evaluate the contribution from each interval. Employing m intervals of length $\ell = (b-a)/m$, we get

$$\int_a^b f(x)dx = \sum_{j=1}^m \int_{a+(j-1)\ell}^{a+j\ell} f(x)dx$$

Each of the integrals in the summation can be transformed to have limits -1 to 1 by taking

$$x = \alpha_j + \beta t$$

with

$$\alpha_j = a + (j - .5)\ell \qquad \beta = .5\ell$$

Therefore we obtain the identity

$$\int_a^b f(x)dx = \sum_{j=1}^m .5\ell \int_{-1}^1 f(\alpha_j + \beta t)dt$$

Applying the same n-point quadrature formula in each of m equal intervals gives what is termed a composite formula

$$\int_a^b f(x)dx = .5\ell \sum_{j=1}^m \sum_{i=1}^n W_i f(\alpha_j + \beta x_i) + E$$

By interchanging the summation order in the previous equation we get

$$\int_a^b f(x)dx = .5\ell \sum_{i=1}^n W_i \sum_{j=1}^m f(\alpha_j + \beta x_i) + E$$

Let us now turn to certain choices of weight factors and base points. Two of the most widely used approximations assume that the integrand is represented satisfactorily as either piecewise linear or piecewise cubic. Approximating the integrand by a straight line through the integrand end points gives the following formula

$$\int_{-1}^1 f(x)dx = f(-1) + f(1) + E$$

A much more accurate formula results by using a cubic approximation matching the integrand at $x = -1, 0, 1$. Let us write

$$f(x) = c_1 + c_2 x + c_3 x^2 + c_4 x^3$$

then

$$\int_{-1}^1 f(x)dx = 2c_1 + \frac{2}{3}c_3$$

Evidently the linear and cubic terms do not influence the integral value. Also, $c_1 = f(0)$ and $f(-1) + f(1) = 2\, c_1 + 2\, c_3$ so that

$$\int_{-1}^1 f(x)dx = \frac{1}{3}[f(-1) + 4f(0) + f(1)] + E$$

The error E in this formula is zero when the integrand is any polynomial of order 3 or lower. Expressed in terms of more general limits this result is

$$\int_a^b f(x)dx = \frac{(b-a)}{6}[f(a) + 4f(.5\,(a+b)) + f(b)] + E$$

which is called Simpson's rule.

Analyzing the integration error for a particular choice of integrand and quadrature formula can be complex. In practice, the usual procedure taken is to apply a composite formula with m chosen large enough so the integration error is expected to be negligibly small. The value for m is then increased until no further significant change in the integral approximation results. Although this procedure involves some risk of error, adequate results can be obtained in most practical situations.

In the subsequent discussions the integration error which results by replacing an integral by a weighted sum of integrand values will be neglected. It must nevertheless be kept in mind that this error depends on the base points, weight factors, and the particular integrand. Most importantly, the error typically decreases as the number of function values is increased.

It is convenient to summarize the composite formulas obtained by employing a piecewise linear or piecewise cubic integrand approximation. Using m intervals and letting $\ell = (b - a)/m$, it is easy to obtain the composite trapezoidal formula which is

$$\int_a^b f(x)dx = \ell \left[\frac{f(a) + f(b)}{2} + \sum_{j=1}^{m-1} f(a + j\ell) \right]$$

This formula assumes that the integrand is satisfactorily approximated as piecewise linear. A similar but much more accurate result is obtained for the composite integration formula based on cubic approximation. For this case, taking m intervals implies $2m + 1$ function evaluations. If we let $g = (b - a)/(2m)$ and $h = 2g$, then

$$x_j = a + gj \qquad f_j = f(x_j) \qquad j = 0, 1, 2, \ldots, 2m$$

where $f(x_0) = f(a)$ and $f(x_{2m}) = f(b)$. Combining results for all intervals gives

$$\int_a^b f(x)dx = \frac{h}{6}[(f_0 + 4f_1 + f_2) + (f_2 + 4f_3 + f_4) + \ldots +$$

$$(f_{2m-2} + 4f_{2m-1} + f_{2m})]$$

The terms in the previous formula can be rearranged into a form more convenient for computation. We get

$$\int_a^b f(x)dx = \frac{h}{6} \left[f(a) + 4f_1 + f(b) + \sum_{i=1}^{m-1} (4f_{2i+1} + 2f_{2i}) \right]$$

This formula, known as the composite Simpson rule, is one of the most commonly used numerical integration methods. A MATLAB implementation of Simpson's rule is listed below.

```
function ansr=simpson(funct,a,b,neven)
%
% ansr=simpson(funct,a,b,neven)
% ~~~~~~~~~~~~~~~~~~~~~~~~~~~~~~
%
% This function integrates "funct" from
% "a" to "b" by Simpson's rule using
% "neven+1" function values.  Parameter
% "neven" should be an even integer.
%
% Example use:  ansr=simpson('sin',0,pi/2,4)
%
% funct    -  character string name of
%             function integrated
% a,b      -  integration limits
% neven    -  an even integer defining the
%             number of integration intervals
% ansr     -  Simpson rule estimate of the
%             integral
%
% User m functions called: argument funct
%-----------------------------------------------

ne=max(2,2*round(.1+neven/2)); d=(b-a)/ne;
x=a+d*(0:ne); y=feval(funct,x);
ansr=(d/3)*(y(1)+y(ne+1)+4*sum(y(2:2:ne))+...
    2*sum(y(3:2:ne-1)));
```

An important goal in numerical integration is to achieve accurate results with only a few function evaluations. It was shown
for Simpson's rule that three function evaluations are enough to
exactly integrate a cubic polynomial. By choosing the base point
locations properly, a much higher accuracy can be achieved for a
given number of function evaluations than would be obtained by
using evenly spaced base points. Results from orthogonal function
theory lead to the following conclusions. If the base points are
located at the zeros of the Legendre polynomials (all these zeros

are between -1 and 1) and the weight factors are computed as certain functions of the base points, then the formula

$$\int_{-1}^{1} f(x)dx = \sum_{i=1}^{n} W_i f(x_i)$$

is exact for a polynomial integrand of degree $2n - 1$. Although the theory proving this property is not elementary, the final results are quite simple. The base points and weight factors for a particular order can be computed once and stored as data statements in the program. Formulas which use the Legendre polynomial roots as base points are called Gauss quadrature formulas. In a typical application, Gauss integration gives much more accurate results than Simpson's rule for an equivalent number of function evaluations. Since it is equally easy to use, the Gauss formula is preferable to the well known Simpson's rule.

MATLAB also provides two functions **quad** and **quad8** which perform numerical integration by adaptive methods. These methods repeatedly modify estimates of an integral until

$$abs \left(\frac{\text{error\_estimate}}{\text{integral\_value}} \right)$$

is smaller than some error tolerance. Presently, such procedures have not yet been suitably refined to produce accuracy levels comparable to what is routinely taken for granted (for example, in the evaluation of elementary functions). Since the adaptive calculation process is vulnerable to convergence failure, error estimates are usually made quite conservatively in order to produce results at least as good as what the chosen error tolerances would imply. The MATLAB functions **quad** and **quad8** usually work well. However, these integrators should be used with some degree-of-caution. These functions employ conservative error measures, so warning messages sometimes result when simple functions such as $\sqrt{x}$ are integrated. Furthermore, choosing too crude an error tolerance may yield inaccurate results; whereas, using too stringent a tolerance can require excessive computation. If a user is sufficiently informed to choose a nonadaptive integrator known to be accurate enough for a particular application, that formula may frequently run much faster than a corresponding adaptive formula. For example, evaluating $\int_0^1 e^{10x} \cos(10\pi x)\, dx$ with **quad8** and a tolerance of 10^{-3} takes 14 times as much computer time and produces an integration error 127 times as large as that resulting when **gquad10** (a tenth-order Gauss integrator provided by the authors) is used with 40 function evaluations.

5.2 Concepts of Gauss Integration

To explain the process followed to obtain base points and weight factors in a Gauss quadrature formula, several properties of Legendre polynomials are summarized. Additional details on their properties can be found in the handbook by Abramowitz and Stegun [1].

The Legendre polynomial $P_n(x)$ is of degree n and has n distinct zeros in the interval $-1 < x < 1$. Although a general polynomial expression can be derived for $P_n(x)$, it is most conveniently computed recursively from the formulas

$$P_0(x) = 1 \qquad P_1(x) = x$$

$$P_n(x) = \frac{n-1}{n}[xP_{n-1}(x) - P_{n-2}(x)] + xP_{n-1}(x) \qquad n \geq 2$$

Derivatives of $P_n(x)$ can also be generated recursively from

$$P_n'(x) = \frac{n[P_{n-1}(x) - xP_n(x)]}{1 - x^2}$$

The Legendre polynomial is one of the fundamental solutions to Legendre's differential equation

$$(1 - x^2)y''(x) - 2xy'(x) + n(n+1)y = 0$$

This differential equation can be used to show that the polynomials of different orders satisfy an orthogonality relation which is

$$\int_{-1}^{1} P_n(x)P_m(x)\, dx = 0 \qquad n \neq m$$

$$\int_{-1}^{1} P_n^2(x) = \frac{2}{2n+1}$$

The polynomials can also be represented as the coefficients in the power series expansion of the related generating function

$$\frac{1}{\sqrt{1 - 2xt + t^2}} = \sum_{n=0}^{\infty} P_n(x)t^n$$

By setting $t = e^{i\theta}$ and using the Fourier series techniques described in Chapter 6, an integral representation can be obtained of the form

$$P_n(x) = \frac{\sqrt{2}}{\pi} \int_0^{\arccos(x)} \left(\frac{\cos(n + \frac{1}{2})\varphi \, d\varphi}{\sqrt{\cos\varphi - \cos\theta}} \right)$$

Next we describe the algorithm used by Davis and Rabinowitz [22] for computing Gauss base points and weight factors. The base points are located at the zeros of $P_n(x)$ and the weight factors are simple functions of those zeros.

Since it is desirable to mix MATLAB and Legendre function notation in the following discussion, we let

$$\begin{aligned}
P_n(x) &\iff \text{P(n,x)} \\
P_n'(x) &\iff \text{Pd1(n,x)} \\
P_n''(x) &\iff \text{Pd2(n,x)} \\
\text{Xk} &\iff \text{k'th iteration for vector of roots} \\
\text{Xk1} &\iff \text{(k+1)'th iteration for vector of roots}
\end{aligned}$$

It can be shown that once a vector X containing the base points has been computed, the vector of weight factors is

```
w=2*(1-x.*x)./(n*P(n-1,x)).^2
```

Furthermore, asymptotic expansions for the Legendre polynomials lead to the following initial root estimates needed to start a Newton method iteration for root refinement.

```
m=fix((n+1)/2); nn=(1-(1-1/n))/(8*n*n);
x0=nn*cos(pi*(3:4:mm))/(4*n+2);
```

The iteration formula for root refinement is

```
Xk1=Xk-(P(n,Xk)./Pd1(n,Xk)).*(1+ ...
    P(n,Xk).*Pd2(n,Xk)/.(2*Pd1(n,Xk).^2))
```

The closeness of the initial root estimates and the accuracy of the iteration formula leads to remarkably fast convergence. Double precision accuracy is achieved after two iterations. All of the necessary function values and derivatives are obtained with the aid of recursion formulas. The procedure is fast enough that computing base points and weight factors for a ten point formula only requires about 0.1 second on a modern microcomputer. Alternatively, the base points and weights of a particular order can simply be computed once and stored as data in an integration formula. Given below are MATLAB functions **grule** and **gquad** which are vectorized translations of similar routines from Davis and Rabinowitz [22]. Also presented are functions **gquad6** and **gquad10** containing embedded base points and weight factors. These functions are used in a test program called **runcases**. This program is used to compare results which **quad**, **quad8**, and **gquad10** provide for several difficult test cases. Table 5.1 summarizes output from **runcases**.

5.3   Examples Comparing Different Integration Methods

We conclude this section by comparing answers obtained when **quad**, **quad8**, and **gquad10** are used to integrate several difficult integrands over limits of zero to one. The examples include a) $\sqrt{x}$, which has infinite slope at $x = 0$; b) $\log(x)$, which is infinite at $x = 0$ but is still integrable; c) **humps**(x), which is a test function provided in MATLAB and has complex poles near $x = 0.3$ and $x = 0.9$, d) $e^{10x} \cos(10\pi x)$, and e) is a highly oscillatory function, the integral of which gives the Bessel function $J_{20}(20)$. Table 5.1 summarizes results obtained and shows function value counts, cpu times, error percentages, and relative computation times (compared with **gquad10**).

Table 5.1 indicates that **quad** gave numerous singularity warnings for **sqrt(x)**. Both **quad** and **quad8** failed on **log(x)**. The integrator **gquad10** gave good results on all problems and was between 5 and 140 times faster than the other integrators. The authors have found high order Gauss quadrature to be a useful tool in many applications. Readers may wish to consider Gauss quadrature as a useful alternative to **quad** and **quad8**.

Function	Method	# of function values	cpu seconds	percent error	t/tgquad10	Notes
$sqrt(x)$	quad	80	0.604	6E-4	27.5	1
	quad8	177	0.406	4E-6	18.5	
	gquad10	100	0.022	4E-4	1	
$log(x)$	quad	16	0.066	∞	-	2
	quad8	17	0.044	∞	-	2
	gquad10	100	0.022	6E-2	1	
$humps(x)$	quad	120	0.648	4E-5	20.3	
	quad8	65	0.154	5E-5	4.8	
	gquad10	100	0.032	4E-12	1	
$\exp(10x)\cos(10\pi x)$	quad	492	2.540	2E-4	75	
	quad8	129	0.286	8E-8	8.4	
	gquad10	200	0.034	2E-13	1	
$\cos(20\pi x - 20\sin(\pi x))$	quad	1100	6.150	2E-5	140	
	quad8	289	0.660	4E-6	15	
	gquad10	200	0.044	2E-13	1	
Note 1: Singularity warnings issued						
Note 2: Error indicating $\log(0)$ encountered						

TABLE 5.1. Comparative Results from Three Integrators

5.3.1   EXAMPLE: COMPUTATION OF BASE POINTS AND WEIGHT FACTORS

5.3.1.1   Script File runcases

```
 1: % Example:  runcases
 2: % ~~~~~~~~~~~~~~~~~~~
 3: % Comparison of numerical integration results
 4: % produced by quad, quad8, and gquad10. The
 5: % functions integrated over limits zero to
 6: % one are: sqrt(x), log(x), humps(x),
 7: % exp(10*x)*cos(10*x), and
 8: % cos(20*pi*x+20*sin(pi*x)). Successive rows
 9: % in matrix r contain the following values.
10: %
11: % row            result
12: % 1         exact integral values
13: % 2         cputime using quad
14: % 3         cputime using quad8
15: % 4         cputime using gquad10
16: % 5         time_quad./time_gquad10
17: % 6         time_quad8./time_gquad10
18: % 7         integrals using quad
19: % 8         integrals using quad8
20: % 9         integrals using gquad10
21: % 10        number of function values by quad
22: % 11        number of function values by quad8
23: % 12        number of function values by gquad10
24: % 13        percent error using quad
25: % 14        percent error using quad8
26: % 15        percent error using gquad10
27: %
28: % User m functions called:
29: %    besf, expc, hmpf, logf, sqtf,
30: %    gquad10
31: %-----------------------------------------------
32:
33: clear
34: global nval_
35:
36: % results are saved in matrix r
37: r=zeros(15,5); ng10=[10,10,10,20,20]; m=5;
38:
39: % names of functions integrated for 0 to 1
```

```
40: a=['sqtf';'logf';'hmpf';'expc';'besf'];
41:
42: % exact values of the integrals
43: r(1,:)=[2/3, -1, gquad10('hmpf',0,1,100), ...
44:     real((exp(10+10*pi*i)-1)/(10+10*pi*i)),...
45:     bessel(20,20)];
46:
47: % integrate using quad, quad8 and gquad10
48: for k=1:5
49:   ak=['''',a(k,:),''''];  nk=num2str(ng10(k));
50:   str1=['r(7,k)=quad(',ak,',0,1);'];
51:   str2=['r(8,k)=quad8(',ak,',0,1);'];
52:   str3=['r(9,k)=gquad10(',ak,',0,1,',nk,');'];
53:
54: % integrate using quad
55:   nval_=0;  r(2,k)=cputime;
56:     for j=1:m, eval(str1); end
57:   r(2,k)=cputime-r(2,k);  r(10,k)=nval_;
58:
59: % integrate using quad8
60:   nval_=0;  r(3,k)=cputime;
61:     for j=1:m, eval(str2); end
62:   r(3,k)=cputime-r(3,k);  r(11,k)=nval_;
63:
64: % integrate using gquad10
65:   nval_=0;  r(4,k)=cputime;
66:     for j=1:m, eval(str3); end
67:   r(4,k)=cputime-r(4,k);  r(12,k)=nval_;
68: end
69:
70: % time ratios relative to gquad10
71: r(5,:)=r(2,:)./r(4,:);  r(6,:)=r(3,:)./r(4,:);
72:
73: % percent errors of different methods
74: r(13,:)=100*abs(1-r(7,:)./r(1,:));
75: r(14,:)=100*abs(1-r(8,:)./r(1,:));
76: r(15,:)=100*abs(1-r(9,:)./r(1,:));
77:
78: exaval=r(1,:);tquad=r(2,:)/m;tquad8=r(3,:)/m;
79: tgqd10=r(4,:)/m;quadrat=r(5,:);quad8rat=r(6,:);
80: qdval=r(7,:);qd8val=r(8,:);gqd10vl=r(9,:);
81: qdnf=r(10,:)/m;qd8nf=r(11,:)/m;gqd10nf=r(12,:)/m;
82: erqd=r(13,:);erqd8=r(14,:);ergqd10=r(15,:);
83:
84: % Print results obtained from the test cases
```

```
85:
86: format long
87: exaval,
88: qdval,    qd8val,    gqd10vl
89: erqd,     erqd8,     ergqd10
90: tquad,    tquad8,    tgqd10
91: qdnf,     qd8nf,     gqd10nf
92: quadrat,  quad8rat
93: format short
94:
95: disp('All Done')
```

5.3.1.2   Function besf

```
1: function y=besf(x)
2: %
3: % y=besf(x)
4: % ~~~~~~~~~
5: %
6: % Companion function for script runcases.
7: %
8: % User m functions called:  none
9: %------------------------------------------------
10:
11: global nval_
12: nval_=nval_+length(x);
13: % integrand defining bessel(20,20)
14: y=cos(20*pi*x-20*sin(pi*x));
```

5.3.1.3   Function expc

```
1: function y=expc(x)
2: %
3: % y=expc(x)
4: % ~~~~~~~~~
5: %
6: % Companion function for script runcases.
7: %
8: % User m functions called:  none
9: %------------------------------------------------
10:
```

```
11: global nval_
12: nval_=nval_+length(x);
13: y=exp(10*x).*cos(10*pi*x);
```

5.3.1.4 Function hmpf

```
1: function y = hmpf(x)
2: %
3: % y = hmpf(x)
4: %  ~~~~~~~~~~~
5: %
6: % Companion function for script runcases.
7: %
8: % User m functions called:  none
9: %-----------------------------------------------
10:
11: global nval_
12: nval_=nval_+length(x);
13: y=1 ./((x-.3).^2+.01)+1 ./((x-.9).^2+.04)-6;
```

5.3.1.5 Function logf

```
1: function y=logf(x)
2: %
3: % y=logf(x)
4: %  ~~~~~~~~~
5: %
6: % Companion function for script runcases.
7: %
8: % User m functions called:  none
9: %-----------------------------------------------
10:
11: global nval_
12: nval_=nval_+length(x);
13: y=log(x);
```

5.3.1.6 Function sqtf

```
1: function y=sqtf(x)
2: %
```

```
 3: % y=sqtf(x)
 4: % ~~~~~~~~~
 5: % Companion function for script runcases.
 6: %
 7: % User m functions called:  none
 8: %-------------------------------------------------
 9:
10: global nval_
11: nval_=nval_+length(x);
12: y=sqrt(x);
```

5.3.1.7   Function grule

```
 1: function [bp,wf]=grule(n)
 2: %
 3: % [bp,wf]=grule(n)
 4: % ~~~~~~~~~~~~~~~~
 5: % This function computes Gauss base points and
 6: % weight factors using the algorithm given by
 7: % Davis and Rabinowitz in 'Methods of
 8: % Numerical Integration', page 369,
 9: % Academic Press, 1975.
10: %
11: % n       - order of Gauss quadrature used
12: % bp,wf   - vectors containing base points and
13: %           weight factors
14: %
15: % User m functions called:  none.
16: %-------------------------------------------------
17:
18: bp=zeros(n,1); wf=bp; iter=2;
19: m=fix((n+1)/2); e1=n*(n+1);
20: mm=4*m-1; t=(pi/(4*n+2))*(3:4:mm);
21: nn=(1-(1-1/n)/(8*n*n));
22: xo=nn*cos(t);
23: for j=1:iter
24:    pkm1=1; pk=xo;
25:    for k=2:n
26:        t1=xo.*pk; pkp1=t1-pkm1-(t1-pkm1)/k+t1;
27:        pkm1=pk; pk=pkp1;
28:    end
29:    den=1.-xo.*xo;
30:    d1=n*(pkm1-xo.*pk); dpn=d1./den;
```

```
31:  d2pn=(2.*xo.*dpn-e1.*pk)./den;
32:  d3pn=(4*xo.*d2pn+(2-e1).*dpn)./den;
33:  d4pn=(6*xo.*d3pn+(6-e1).*d2pn)./den;
34:  u=pk./dpn; v=d2pn./dpn;
35:  h=-u.*(1+(.5*u).* ...
36:    (v+u.*(v.*v-u.*d3pn./(3*dpn))));
37:  p=pk+h.*(dpn+(.5*h).*(d2pn+(h/3).* ...
38:    (d3pn+.25*h.*d4pn)));
39:  dp=dpn+h.*(d2pn+(.5*h).* ...
40:    (d3pn+h.*d4pn/3));
41:  h=h-p./dp; xo=xo+h;
42: end
43: bp=-xo-h;
44: fx=d1-h.*e1.*(pk+(h/2).*(dpn+(h/3).*(...
45:    d2pn+(h/4).*(d3pn+(.2*h).*d4pn))));
46: wf=2*(1-bp.^2)./(fx.*fx);
47: if (m+m) > n, bp(m)=0; end
48: if ~((m+m) == n), m=m-1; end
49: jj=1:m; n1j=(n+1-jj);
50: bp(n1j)=-bp(jj); wf(n1j)=wf(jj);
```

5.3.1.8 Function gquad

```
1: function area= ...
2:           gquad(fun,xlow,xhigh,mparts,bp,wf)
3: %
4: % area = gquad (fun,xlow,xhigh,mparts,bp,wf)
5: % ~~~~~~~~~~~~~~~~~~~~~~~~~~~~~~~~~~~~~~~~~~~~
6: % This function evaluates the integral of an
7: % externally defined function fun(x) between
8: % limits xlow and xhigh. The numerical
9: % integration is performed using a composite
10: % Gauss integration rule.  The whole interval
11: % is divided into mparts subintervals and the
12: % integration over each subinterval is done
13: % with an nquad point Gauss formula which
14: % involves base points bp and weight factors
15: % wf.  The normalized interval of integration
16: % for the bp and wf constants is -1 to +1.
17: % The algorithm is described by the summation
18: % relation
19: %
20: % x=b
```

```
21: % integral(f(x)*dx)=
22: %  x=a
23: %
24: %               j=n k=m
25: %        d1*sum sum(wf(j)*fun(a1+d*k+d1*bp(j)))
26: %               j=1 k=1
27: %
28: % where bp are base points, wf are
29: % weight factors
30: %        m = mparts, and n = length(bp) and
31: %        d = (b-a)/m, d1 = d/2, a1 = a-d1
32: %
33: % The base points and weight factors must
34: % first be generated by a call to grule of
35: % the form
36: %
37: % [bp,wf] = grule(nquad)
38: %
39: % fun          - function to be integrated
40: % xlow,xhigh   - integration limits
41: % mparts       - number of integration intervals
42: %                used
43: % bp,wf        - base points and weight factors
44: %                obtained by an initial call
45: %                to grule
46: % area         - numerically approximated
47: %                integral value
48: %
49: % User m functions called:  argument fun
50: %-------------------------------------------------
51:
52: bp=bp(:); wf=wf(:);
53: d=(xhigh-xlow)/mparts; d2=d/2;
54: nquad=length(bp);
55: x=(d2*bp)*ones(1,mparts)+ ...
56:   (d*ones(nquad,1))*(1:mparts);
57: x=x(:)+(xlow-d2);
58: fv=feval(fun,x);
59: wv=wf*ones(1,mparts);
60: area=d2*(wv(:)'*fv(:));
```

144

5.3.1.9 Function gquad6

```
 1: function area=gquad6(fun,xlow,xhigh,mparts)
 2: %
 3: % area = gquad6(fun,xlow,xhigh,mparts)
 4: % ~~~~~~~~~~~~~~~~~~~~~~~~~~~~~~~~~~~~~~~~
 5: % This function determines the area under an
 6: % externally defined function fun(x) between
 7: % limits xlow and xhigh. The numerical
 8: % integration is performed using a composite
 9: % gauss integration rule. The whole interval
10: % is divided into mparts subintervals and the
11: % integration over each subinterval is done
12: % with a six point Gauss formula which
13: % involves base points bp and weight factors
14: % wf. The normalized interval of integration
15: % for the bp and wf constants is -1 to +1.
16: % The algorithm is structured in terms of a
17: % parameter mquad = 6 which can be changed
18: % along with bp and wf to accommodate a
19: % different order formula. The composite
20: % algorithm is described by the following
21: % summation relation:
22: %
23: % x=b
24: % integral( f(x)*dx ) =
25: % x=a
26: %
27: %         j=n k=m
28: %     d1*sum sum( wf(j)*fun(a1+d*k+d1*bp(j)) )
29: %         j=1 k=1
30: %
31: % where d = (b-a)/m, d1 = d/2, a1 = a-d1,
32: %       m = mparts, and n = nquad.
33: %
34: % User m functions called:  argument fun
35: %-------------------------------------------------
36:
37: % The weight factors are
38: wf=[ 1.71324492379170d-01; ...
39:      3.60761573048139d-01;...
40:      4.67913934572691d-01];
41: wf=[wf;wf([3 2 1])];
42:
```

```
43: % The base points are
44: bp=[-9.32469514203152d-01; ...
45:     -6.61209386466265d-01;...
46:     -2.38619186083197d-01];
47: bp=[bp;-bp([3 2 1])];
48:
49: d=(xhigh-xlow)/mparts;
50: d2=d/2; nquad=length(bp);
51: x=(d2*bp)*ones(1,mparts)+ ...
52:    (d*ones(nquad,1))*(1:mparts);
53: x=x(:)+(xlow-d2);
54: fv=feval(fun,x); wv=wf*ones(1,mparts);
55:
56: area=d2*(wv(:)'*fv(:));
```

5.3.1.10   Function gquad10

```
 1: function area=gquad10(fun,xlow,xhigh,mparts)
 2: %
 3: % area = gquad10(fun,xlow,xhigh,mparts)
 4: % ~~~~~~~~~~~~~~~~~~~~~~~~~~~~~~~~~~~~~
 5: % This function determines the area under an
 6: % externally defined function fun(x) between
 7: % limits xlow and xhigh. The numerical
 8: % integration is performed using a composite
 9: % gauss integration rule. The whole interval
10: % is divided into mparts subintervals and the
11: % integration over each subinterval is done
12: % with a ten point Gauss formula which
13: % involves base points bp and weight factors
14: % wf. The normalized interval of integration
15: % for the bp and wf constants is -1 to +1.
16: % The algorithm is structured in terms of a
17: % parameter mquad = 6 which can be changed
18: % along with bp and wf to accommodate a
19: % different order formula. The composite
20: % algorithm is described by the following
21: % summation relation:
22: %
23: % x=b
24: % integral( f(x)*dx ) =
25: % x=a
26: %
```

```
27: %              j=n k=m
28: %           d1*sum sum( wf(j)*fun(a1+d*k+d1*bp(j)) )
29: %              j=1 k=1
30: %
31: % where d = (b-a)/m, d1 = d/2, a1 = a-d1,
32: %           m = mparts, and n = nquad.
33: %
34: % User m functions called:  argument fun
35: %-----------------------------------------------
36:
37: if nargin ==3, mparts=1; end
38:
39: % The weight factors are
40: wf=[  6.66713443086879e-02; ...
41:      14.94513491505806e-02; ...
42:      21.90863625159821e-02; ...
43:      26.92667193099964e-02; ...
44:      29.55242247147528e-02 ];
45: wf=[wf; wf([5:-1:1])];
46:
47: % The base points are
48: bp=[-97.39065285171718e-02; ...
49:     -86.50633666889846e-02; ...
50:     -67.94095682990245e-02; ...
51:     -43.33953941292472e-02; ...
52:     -14.88743389816312e-02 ];
53: bp=[bp;-bp([5:-1:1])];
54:
55: d=(xhigh-xlow)/mparts;
56: d2=d/2; nquad=length(bp);
57: x=(d2*bp)*ones(1,mparts)+ ...
58:   (d*ones(nquad,1))*(1:mparts);
59: x=x(:)+(xlow-d2);
60: fv=feval(fun,x); wv=wf*ones(1,mparts);
61:
62: area=d2*(wv(:)'*fv(:));
```

5.4   Line Integrals for Geometric Properties of Plane Areas

The next article applies Gauss quadrature and spline interpolation to analyze a practical problem. Engineering applications often require computation of geometrical properties of plane areas. The main properties needed are the area,

$$A = \int\int dx\, dy$$

the first moments of area,

$$A_x = \int\int x\, dx\, dy \qquad A_y = \int\int y\, dx\, dy$$

and the three inertial moments.

$$A_{xx} = \int\int x^2\, dx\, dy \qquad A_{xy} = \int\int x\, y\, dx\, dy$$

$$A_{yy} = \int\int y^2\, dx\, dy$$

This problem is important enough that we will analyze it for regions having one or several parts which can be solid or can contain holes. Two types of solutions are presented. The first handles any region bounded by straight lines and circular arcs. The second solution deals with more general shapes where the boundary is approximated by a spline curve.

The desired area properties all have the form

$$I_{nm} = \int\int x^n y^m\, dx\, dy$$

where n and m are integers. This two-dimensional integral can be converted into a one-dimensional integral evaluated over the boundary curve.[1] The one-dimensional integrals can be either computed exactly or numerically. Consider the general area in Figure 5.1 bounded by a curve L described in parametric form as

$$(x(t), y(t)) \qquad a \le t \le b$$

---

[1]Wilson and Farrior [94] published an early paper on this topic.

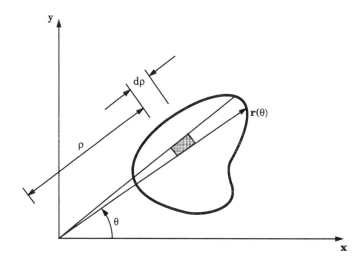

FIGURE 5.1. General Two Dimensional Area

The area integral can be transformed into a line integral by using polar coordinates in the form

$$x = \rho \cos \theta \qquad y = \rho \sin \theta$$

$$\rho = (x^2 + y^2)^{\frac{1}{2}} \qquad \theta = \tan^{-1}\left(\frac{y}{x}\right)$$

$$dA = \rho \, d\rho \, d\theta \qquad d\theta = \frac{x \, dy - y \, dx}{x^2 + y^2}$$

These relations yield

$$I_{nm} = \int_{\theta} \int_{0}^{r(\theta)} (\rho \cos \theta)^n (\rho \sin \theta)^m \, \rho \, d\rho \, d\theta$$

$$I_{nm} = \int_{\theta} (\cos \theta)^n (\sin \theta)^m \int_{0}^{r} \rho^{n+m+1} d\rho \, d\theta$$

$$I_{nm} = \frac{1}{n+m+2} \int_{\theta} [\rho \cos \theta]^n [\rho \sin \theta]^m \, \rho^2 d\theta$$

Returning to cartesian coordinates gives

$$\int \int x^n y^m dx \, dy = \frac{1}{n+m+2} \int_{L} x^n y^m \, (x \, dy - y \, dx)$$

This formula is good for non-negative integer values of n and m. The formula even works for negative indices provided $x = 0$ is outside L for negative n, and $y = 0$ is outside L for negative m. The important cases include:

- $n = m = 0$ for area,

- $n = 1$, $m = 0$ or $n = 0$, $m = 1$ for first moments, and

- $n = 2$, $m = 0$, or $n = m = 1$, or $n = 0$, $m = 2$ for inertial moments.

A general boundary shape may have several parts each of which can be represented by a parametric equation. For example, a typical line segment from (x_1, y_1) to (x_2, y_2) can be expressed as

$$x = x_1 + (x_2 - x_1)t \qquad y = y_1 + (y_2 - y_1)t \qquad 0 \le t \le 1$$

Since

$$x \, dy - y \, dx = (x_1 y_2 - x_2 y_1)dt = g_{12}dt$$

the integral contributions from the line segment are

$$\frac{1}{2} \int x \, dy - y \, dx = \frac{1}{2}(x_1 y_1 - y_1 x_2) = \frac{1}{2}g_{12}$$

$$\frac{1}{3} \int x(x \, dy - y \, dx) = \frac{1}{6}(x_1 + x_2)g_{12}$$

$$\frac{1}{3} \int y(x \, dy - y \, dx) = \frac{1}{6}(y_1 + y_2)g_{12}$$

$$\frac{1}{4} \int x^2(x \, dy - y \, dx) = \frac{1}{12}(x_1^2 + x_1 x_2 + x_2^2)g_{12}$$

$$\frac{1}{4} \int xy(x \, dy - y \, dx) = \frac{1}{24}(2x_1 y_1 + x_1 y_2 + x_2 y_1 + 2x_2 y_2)g_{12}$$

$$\frac{1}{4} \int y^2(x \, dy - y \, dx) = \frac{1}{12}(y_1^2 + y_1 y_2 + y_2^2)g_{12}$$

Similar formulas can be developed for a circular arc described as

$$x = x_0 + r \cos t \qquad y = y_0 + r \sin t \qquad \theta_1 \le t \le \theta_2$$

Then

$$x \, dy - y \, dx = r_0 \left[r_0 + x_0 \cos t - y_0 \sin t \right] dt$$

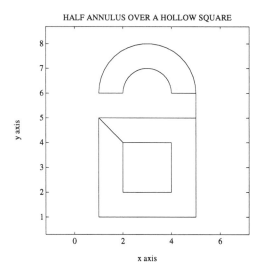

FIGURE 5.2. Half Annulus Over a Hollow Square

and the integral pertaining to area is

$$\frac{1}{2}\int (x\ dy - y\ dx) = r_0 \left[r_0 \langle \theta_2 - \theta_1 \rangle + x_0 \langle s_1 - s_2 \rangle + y_0 \langle c_1 - c_2 \rangle \right]$$

where

$$s_1 = \sin\theta_1 \qquad c_1 = \cos\theta_1 \qquad s_2 = \sin\theta_2 \qquad c_2 = \cos\theta_2$$

The other integrals for first and second area moments are similar but algebraically tedious. Those expressions are not repeated here since they are included in the MATLAB functions which compute exactly the properties of any region bounded by straight lines and circular arcs. The realistic geometry described by circular arcs and straight lines in Figure 5.2 involves half of an annulus placed over a hollow square. It will be used later to compare exact results with those obtained when the boundary is approximated by a spline. However, it will be helpful to first consider a simple example to demonstrate some ideas.

5.4.1   GEOMETRY EXAMPLE USING A SIMPLE SPLINE INTERPOLATED BOUNDARY

Let us consider a simple case before proceeding with the treatment of general spline interpolated boundaries. Suppose a set of data points (x_i, y_i), $1 \leq i \leq n$, are to be connected by a smooth closed curve having parametric form $x(t), y(t)$, $a \leq t \leq b$. The area, centroidal coordinates, and inertial moments resulting from evaluation of the six basic integrals are also sought. To get a smooth curve through the data we choose points $[x_n, x_1, x_2, \cdots, x_n, x_1, x_2]$ and $[y_n, y_1, y_2, \cdots, y_n, y_1, y_2]$ which are spline interpolated (using **spline** from MATLAB) as functions of $t = [0, 1, 2, \cdots, n+2]$. The end points added to the data make the curve close and give a smoothly turning tangent. The remaining calculations needed to get area properties are to evaluate $x'(t)$ and $y'(t)$ by a method such as finite differences, and then evaluate the integrals by using an algorithm such as Simpson's rule. The following short program is adequate to handle smooth shapes. An example was run to approximate a rotated ellipse using 20 data points as shown in Figure 5.3. The maximum error in computed geometrical properties was about 0.4%. Although this simple program illustrates the general procedure for computing area properties, dealing with complex shapes having sharp corners takes additional effort. More careful attention is needed regarding boundary slope representation, derivative evaluation, and numerical integration. The general spline approximation problem is discussed next.

152

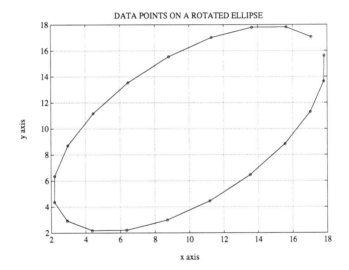

FIGURE 5.3. Rotated Ellipse

MATLAB EXAMPLE

5.4.1.1   Script File elipprop

```
 1: % Example: elipprop
 2: % ~~~~~~~~~~~~~~~~~~~
 3: % Compute the area, centroidal coordinates,
 4: % and inertial moments of a rotated ellipse
 5: % centered at (cx,cy) and having diameters
 6: % of 2*rx and 2*ry
 7: %
 8: % User m functions required:
 9: %    allprop, dife, simpsum
10: %-----------------------------------------------
11:
12: cx=10; cy=10; rx=10; ry=5;
13: np=20; th=2*pi/np*(0:np-1);
14: z=(rx*cos(th)+i*ry*sin(th))* ...
15:    exp(i*pi/4)+(cx+i*cy);
16: xdat=real(z); ydat=imag(z);
17:
18: clf; plot(xdat,ydat,'w',xdat,ydat,'ow')
19: xlabel('x axis'), ylabel('y axis'), grid
20: title('DATA POINTS ON A ROTATED ELLIPSE');
21: disp(' '); disp('Press RETURN to continue');
22: disp(' '); pause
23:
24: [area,xb,yb,axx,axy,ayy]= ...
25:    allprop(xdat,ydat,101);
26:
27: fprintf('area   = %7.4f\n',area)
28: fprintf('xcentr = %7.4f\n',xb)
29: fprintf('ycentr = %7.4f\n',yb)
30: fprintf('axx    = %7.4f\n',axx)
31: fprintf('axy    = %7.4f\n',axy)
32: fprintf('ayy    = %7.4f\n',ayy)
33:
34: % Exact values would give unity
35: % for aa, xx and yy
36: aa=area/(pi*rx*ry); xx=xb/cx; yy=yb/cy;
37:
38: fprintf(...
39:   '\napprox_area / exact_area   = %7.4f\n',aa)
40: fprintf(...
```

```
41:  '\napprox_xcentr / exact_xcntr = %7.4f\n',xx)
42: fprintf(...
43:  '\napprox_ycentr / exact_ycntr = %7.4f\n',yy)
44: disp(' '), disp('All Done')
```

5.4.1.2 Function dife

```
1: function ydif=dife(y,h)
2: %
3: % ydif=dife(y,h)
4: % ~~~~~~~~~~~~~~
5: % This function differentiates data evenly
6: % spaced with the argument increment being h.
7: %
8: % User m functions called:  none
9: %---------------------------------------------
10:
11: % Use central differences for interior points
12: y=y(:); n=length(y);
13: ydif=( y(3:n)-y(1:n-2) )/2;
14:
15: % Use forward and backward differences at
16: % the end points
17: ydif=[(-3*y(1)+4*y(2)-y(3))/2; ydif;
18:       (y(n-2)-4*y(n-1)+3*y(n))/2];
19:
20: % Default value of h is one
21: if nargin ==2, ydif=ydif/h; end
```

5.4.1.3 Function simpsum

```
1: function asmp=simpsum(y,a,b)
2: %
3: % asmp=simpsum(y,a,b)
4: % ~~~~~~~~~~~~~~~~~~~
5: % Simpson's rule for a matrix
6: % having an odd number of rows corresponding to
7: % argument values evenly spaced from a to b.
8: %
9: % User m functions called:  none
10: %---------------------------------------------
```

```
11:
12: [n,nc]=size(y); h=(b-a)/(n-1);
13:
14: % If the number of function values is even,
15: % then pad with a zero at the bottom. This
16: % can affect accuracy adversely, so taking n
17: % odd is best.
18: if n==2*fix(n/2), n=n+1; y=[y;0]; b=b+h; end
19:
20: % Accumulate the summation
21: asmp=(b-a)/(3*(n-1))*(y(1,:)+y(n,:)+...
22:         4*sum(y(2:2:n-1,:))+2*sum(y(3:2:n-2,:)));
```

5.4.1.4   Function allprop

```
 1: function [area,xcentr,ycentr,axx,axy,ayy]=...
 2:                 allprop(xdat,ydat,ntrp)
 3: %
 4: % [area,xcentr,ycentr,axx,axy,ayy]=...
 5: %                 allprop(xdat,ydat,ntrp)
 6: % ~~~~~~~~~~~~~~~~~~~~~~~~~~~~~~~~~~~~~~~~~~~
 7: % This function computes area, centroidal
 8: % coordinates, and second moments of area
 9: % (inertial moments) for a spline curve passed
10: % through points (xdat,ydat).
11: %
12: % area            - area enclosed by the curve
13: % xcentr,ycentr   - centroidal coordinates
14: % axx             - integral( x^2*d(area) )
15: % axy             - integral( y*x*d(area) )
16: % ayy             - integral( y^2*d(area) )
17: %
18: % User m functions called:  dife, simpsum
19: %-------------------------------------------------
20:
21: if nargin < 3, ntrp=101; end; nd=length(xdat);
22: xdat=xdat(:); ydat=ydat(:);
23:
24: % Make curve close and extend the end values
25: xdat=[xdat(nd);xdat;xdat(1:2)];
26: ydat=[ydat(nd);ydat;ydat(1:2)];
27:
28: % Be sure to use an odd number
```

```
29: % of quadrature points
30: if ntrp==2*fix(ntrp/2), ntrp=ntrp+1; end
31: h=(nd-1)/ntrp; t=1:h:nd+1; tdat=0:nd+2;
32:
33: % Spline interpolate to get integrand values
34: x=spline(tdat,xdat,t); x=x(:);
35: y=spline(tdat,ydat,t); y=y(:);
36:
37: % Approximate derivatives by
38: % finite differences
39: xd=dife(x,h); yd=dife(y,h);
40:
41: % Compute area by Simpson's rule
42: a=x.*yd-y.*xd;
43: am=zeros(length(a),6); am(:,1)=a;
44: am(:,2)=a.*x; am(:,3)=a.*y;
45: am(:,4)=am(:,2).*x;
46: am(:,5)=am(:,2).*y; am(:,6)=am(:,3).*y;
47:
48: % Calculate centroidal coordinates
49: v=simpsum(am,1,nd+1); area=v(1)/2;
50: xcentr=v(2)/(3*area); ycentr=v(3)/(3*area);
51: axx=v(4)/4; axy=v(5)/4; ayy=v(6)/4;
```

5.5   Spline Approximation of General Boundary Shapes

In Chapter 4 spline interpolation was used to represent a plane curve by piecewise cubic polynomials. We will briefly review some ideas about splines. Suppose a closed curve is defined by connecting points $(x_1, y_1), \cdots, (x_n, y_n)$ where $x_1 = x_n$, $y_1 = y_n$ is required for the curve to close. At each point (x_j, y_j), a parameter $t_j = j$ is assigned. Functions $x(t)$ and $y(t)$ are determined by spline interpolation using functions **spc** and **spltrp**. Boundary points where slope discontinuities occur, such as the corners of a rectangle, are defined as corner points. The example in Figure 5.4 involving two arcs and ten straight line segments can be approximated well by a spline. Seven points are needed to describe the semicircle accurately. Any spline segment with a corner at each end reduces exactly to a straight line. The boundary curve chosen for spline interpolation has 22 points specified by the point sequence 1-6, 3, 2, 7, 8, 1, 9-22, 9, 1. Furthermore, corners are used at 1-9, 15, 16, and 22. Figure 5.6 displays the merits of corner points by showing the unsatisfactory results produced without the use of corner points. The curve employing corner points, shown in Figure 5.5, approximates well the original geometry involving straight lines and circular arcs.

Let us examine the integrands occurring in the line integrals for area properties of the spline interpolated boundary. Any boundary segment has the form

$$x(t) = a_1 t^3 + a_2 t^2 + a_3 t + a_4 \qquad y(t) = b_1 t^3 + b_2 t^2 + b_3 t + b_4$$

Evidently, $x(t)y'(t) - y(t)x'(t)$ is a polynomial of degree four. For a typical property such as

$$a_{xy} = \frac{1}{4} \int_L xy(x\,dy - y\,dx) = \frac{1}{4} \int_1^n x(t)y(t)[x(t)y'(t) - y(t)x'(t)]dt$$

the integrand has piecewise polynomial form of degree ten. Because an M-point Gauss formula integrates exactly any polynomial of degree $2M - 1$, taking $M = 6$ is adequate to exactly evaluate all the geometrical properties associated with the spline boundary. Furthermore, a boundary with n segments requires a spline interpolation using $n + 1$ points. Thus, $24n$ spline evalua-

158

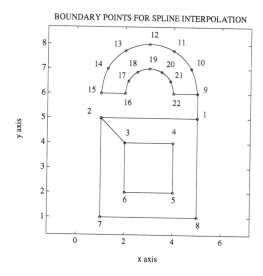

FIGURE 5.4. Boundary Points for Spline Interpolation

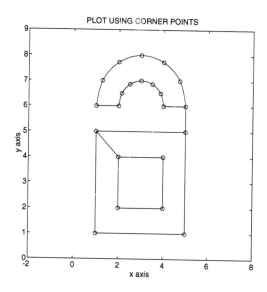

FIGURE 5.5. Plot Using Corner Points

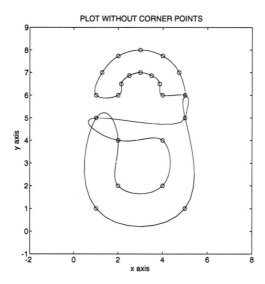

FIGURE 5.6. Plot Without Corner Points

tions are needed to compute x, x', y, and y' values. Using these values to evaluate integrands and accumulate numerical integration sums only requires an additional $120n$ floating point calculations to compute the six area properties of interest.

This chapter concludes with an explanation of the structure of the two geometric property programs. The first program, for geometries defined by circular arcs and straight line segments, evaluates all integrals exactly. The second program, based on spline interpolation, is approximate in the sense that most geometries will not have boundaries exactly defined by a spline. However, no numerical integration error is encountered in the geometric property evaluations because six-point composite Gauss quadrature works exactly for the piecewise polynomial integrands.

Sometimes it is necessary to compute the six area properties for new axes translated and rotated relative to the initial frame. This type of transformation can be described conveniently using complex arithmetic. Let a new frame $\hat{z}$ be obtained by first translating the origin to Z_0 and then rotating the axes by an angle θ counterclockwise. The new axes are related to the initial axes according to

$$\hat{z} = (Z - Z_0)e^{-i\theta} \qquad Z = X + iY \qquad \hat{z} = \hat{x} + i\hat{y}$$

To understand how the area properties transform, start with

$$A = \int\int dx\, dy \qquad AZ = \int\int Z\, dx\, dy \qquad Z_c = \frac{AZ}{A}$$

and introduce

$$P_1 = \frac{1}{2}\int\int Z^2\, dx\, dy \qquad P_2 = \frac{1}{2}\int\int |Z|^2\, dx\, dy$$

which yields the following inertial moment terms.

$$
\begin{aligned}
A_{xx} &= \mathbf{real}(P_1 + P_2) \\
A_{yy} &= \mathbf{real}(P_1 - P_2) \\
A_{xy} &= \mathbf{imag}(P_1)
\end{aligned}
$$

Since P_1 and P_2 concisely describe the inertial properties, let us demonstrate how P_1 and P_2 transform under a translation defined by $z = Z - Z_0$, or

$$z_c = Z_c - Z_0$$

$$p_1 = \frac{1}{2}\int\int z^2\, dx\, dy = P_1 - Z_0 Z_c A + \frac{1}{2}Z_0^2 A$$

$$p_2 = \frac{1}{2}\int\int |z|^2\, dx\, dy = P_2 - \mathbf{real}(\bar{Z}_0 Z_c)A + \frac{1}{2}|Z|_0^2\, A$$

The new inertial moments are

$$\int\int x^2\, dx\, dy = \mathbf{real}(p_1 + p_2) \qquad \int\int y^2\, dx\, dy = \mathbf{real}(p_1 - p_2)$$

$$\int\int xy\, dx\, dy = \mathbf{imag}(p_1)$$

After the translation is completed, the axes can be rotated counterclockwise to a new frame $\hat{z}$ defined by $\hat{z} = ze^{-i\theta}$. The new properties are $\hat{z}_c = z_c e^{-i\theta}$, $\hat{p}_1 = e^{-2i\theta}p_1$, and $\hat{p}_2 = p_1$. Consequently,

$$\int\int \hat{x}^2\, d\hat{x}\, d\hat{y} = \mathbf{real}(\hat{p}_1 + \hat{p}_2)$$

$$\int\int \hat{y}^2\, d\hat{x}\, d\hat{y} = \mathbf{real}(\hat{p}_1 + \hat{p}_2)$$

$$\int \int \hat{x}\hat{y} \ d\hat{x} \ d\hat{y} = \mathbf{imag}(\hat{p}_1)$$

This completes the presentation of the formulas used in function **shftprop** to determine area properties produced by translation and rotation of axes.

5.5.1 Program for Exact Properties of Any Area Bounded by Straight Lines and Circular Arcs

Two programs were written to compute geometrical properties of plane areas. The first program, **arearun**, reads a geometry file specifying the straight lines and circular arcs encountered in a counterclockwise circuit around the boundary. Function **crclovsq** defines the general shape plotted in Figure 5.2. The first data item is a title. Then a matrix named *dat* specifies point data. Nonzero values in the first column of *dat* identify the start of a polyline (connecting several successive points) or an arc. For example, the value 11 indicates a line connecting eleven points with (x, y) coordinates contained in the second and third columns of *dat*. Lines such as 24 and 25 in **crclovsq** specify a single line segment. A value of -2 in column one identifies two data lines characterizing an arc. The center coordinates of the arc are contained in the second two items after the -2. The last two items of the line immediately following this line contain the initial angle and the subtended angle (in radians measured positive counterclockwise). For example, the inner semicircle (lines 34 and 35) starts at *pi* and has a subtended angle of $-pi$ because it turns clockwise. Notice, that symbolic expressions such as *pi*/2 may be used rather than rounded values such as 1.57. Furthermore, it is very important that the figure must close. Otherwise, incorrect results will be produced.

Function **pltundst** produces an undistorted plot of any data values (x, y) and titles the graph with a name designated by *grphnam*. Functions **arcprop**, **lineprop**, and **shftprop** perform geometrical property calculations. Function **areaprop** decodes the data from matrix *dat* and shifts the points by (xs, ys) if required. The data is broken into polylines and arcs. Function **lineprop** evaluates contributions of a polyline. Function **arcprop** produces the contribution from an arc. (Note that the various analytical expressions in this routine are algebraically complex.) Function **areaprop** accumulates the properties and returns a single vector containing the area, the centroidal coordinates, and the three inertial moments. A set of boundary points suitable for plotting is also returned. Function **shftprop**, takes area properties for one axis system and produces values for a new system derived by shifting and rotating the reference axes. Output produced by the program for the data specified by **crclovsq** is used to compute properties for the original axes and for axes shifted to the centroid.

MATLAB EXAMPLE

5.5.1.1   Output from Example arearun

`>> arearun`

```
                GEOMETRICAL PROPERTIES OF PLANE AREAS

 give the script file name which defines the data
 ? > crclovsq

                  SUMMARY OF AREA PROPERTIES
   area      xbar      ybar      axx       axy       ayy
 16.7124   3.0000    4.1251   176.3020  206.8230  359.5365


To compute properties relative to shifted axes input
x,y coordinates of the shifted origin and a rotation
angle measured + counterclockwise.
(Input 0,0,0 to stop)
 ? > 3.0,4.1251,0

                  PROPERTIES FOR SHIFTED AXES
   area      xbar      ybar      axx       axy       ayy
 16.7124   0.0000    0.0000    25.8905   -0.0000   75.1450


To compute properties relative to shifted axes input
x,y coordinates of the shifted origin and a rotation
angle measured + counterclockwise.
(Input 0,0,0 to stop)
 ? > 0,0,0


All done
```

5.5.1.2   Script File crclovsq

```
 1: % Example: crclovsq
 2: % ~~~~~~~~~~~~~~~~~
 3: % Script file defining a half annulus above a
 4: % square which contains a square hole.  Used
 5: % as input for script file arearun.
 6: %
 7: % User m functions required:
 8: %     none
 9: %-------------------------------------------------
10:
11: titl='HALF ANNULUS OVER A HOLLOW SQUARE';
12:
13: % Be sure to traverse the boundary
14: % counterclockwise
15:
16: dat=...
17: [11  ,5  ,5   % <-- polyline traversing hollow
18:  0   ,1  ,5   %      square
19:  0   ,2  ,4
20:  0   ,4  ,4
21:  0   ,4  ,2
22:  0   ,2  ,2
23:  0   ,2  ,4
24:  0   ,1  ,5
25:  0   ,1  ,1
26:  0   ,5  ,1
27:  0   ,5  ,5
28:  2   ,5  ,5   % <-- line joining the square
29:  0   ,5  ,6   %      and half annulus
30: -2   ,3  ,6   % <-- outer semicircle of
31:  0   ,0  ,pi  %      the half annulus
32:  2   ,1  ,6   % <-- left side horizontal line
33:  0   ,2  ,6
34: -1   ,3  ,6   % <-- inner semicircle of
35:  0   ,pi ,-pi%      the half annulus
36:  2   ,4  ,6   % <-- right side horizontal line
37:  0   ,5  ,6
38:  2   ,5  ,6   % <-- final line joining
39:  0   ,5  ,5];%      the annulus and square
```

5.5.1.3   Script File arearun

```
 1: % Example: arearun
 2: % ~~~~~~~~~~~~~~~~~
 3: % This program computes the area, centroidal
 4: % coordinates, moments of inertia, and product
 5: % of inertia for any area bounded by straight
 6: % lines and circular arcs. The boundary
 7: % geometry is also depicted graphically. Each
 8: % piecewise linear boundary part is defined by
 9: % a sequence of x,y coordinates. Each arc is
10: % specified by a radius, center coordinates,
11: % initial polar angle, and a subtended angle.
12: % Computation of properties for new reference
13: % axes translated and rotated relative to the
14: % original reference frame is also provided.
15: %
16: % User m functions required:
17: %    areaprop, shftprop, arcprop, lineprop,
18: %    crclovsq, pltundst, read
19: %-------------------------------------------------
20:
21: fprintf('\n          GEOMETRICAL ')
22: fprintf('PROPERTIES OF PLANE AREAS')
23: fprintf('\n\n give the script file name')
24: fprintf(' which defines the data')
25: fn=input('? > ','s'); eval(fn);
26: [a,x,y]=areaprop(dat);
27: pltundst(x, y,titl);
28: disp(' ');
29: disp('Press RETURN to continue'); pause
30: fprintf('\n                SUMMARY ')
31: fprintf('OF AREA PROPERTIES')
32: fprintf('\n   area         xbar        ')
33: fprintf('ybar        axx')
34: fprintf('        axy         ayy \n')
35: disp(a')
36: while 1
37:    fprintf('\nTo compute properties ')
38:    fprintf('relative to shifted axes input')
39:    fprintf('\nx,y coordinates of shifted ')
40:    fprintf('origin and a rotation angle')
41:    fprintf('\nmeasured + counterclockwise.')
42:    fprintf(' (Input 0,0,0 to stop)')
```

```
43:  [xshf, yshf, theta] = read;
44:  if norm([xshf,yshf,theta])==0, break, end
45:  theta=pi*theta/180;
46:  ashf=shftprop(a,xshf,yshf,theta);
47:  fprintf('\n                    PROPERTIES ')
48:  fprintf('FOR SHIFTED AXES')
49:  fprintf('\n    area        xbar        ')
50:  fprintf('ybar        axx')
51:  fprintf('         axy        ayy \n')
52:  disp(ashf')
53: end
54: fprintf('\n\nAll Done\n')
```

5.5.1.4 Function areaprop

```
1: function [aprops,xb,yb]=areaprop(dat,xs,ys)
2: %
3: % [aprops,xb,yb]=areaprop(dat,xs,ys)
4: % ~~~~~~~~~~~~~~~~~~~~~~~~~~~~~~~~~~~
5: % This function determines geometrical
6: % properties of any area bounded by straight
7: % lines and circular arcs. Coordinates are
8: % shifted by (xs,ys) if these parameters are
9: % included in the argument list.
10: %
11: % dat     - array containing data defining
12: %             arcs and lines
13: % (xs,ys) - coordinates to which origin is
14: %             shifted
15: % aprops  - [area; xbar; ybar; axx; axy; ayy]
16: % xb,yb   - boundary point table suitable for
17: %             plotting
18: %
19: % User m functions called:  arcprop, lineprop
20: %-----------------------------------------------
21:
22: if nargin==1, xs=0; ys=0; end
23: aprops=zeros(6,1); ndx=find(dat(:,1)~=0);
24: jline=1; nparts=length(ndx); xb=[]; yb=[];
25: for k=1:nparts
26:   nk=ndx(k); n=dat(nk,1);
27:   if n>0
28:     jnext=jline+n; kk=jline:jnext-1;
```

```
29:      x=dat(kk,2); y=dat(kk,3); x=x-xs; y=y-ys;
30:      aprops=aprops+lineprop(x,y); jline=jnext;
31:      if nargout >1, xb=[xb;x]; yb=[yb;y]; end
32:    else
33:      rad=-n; r=dat(jline,2:3)-[xs,ys];
34:      th1=dat(jline+1,2);
35:      th12=dat(jline+1,3); jline=jline+2;
36:      aprops=aprops+ ...
37:              arcprop(r(1),r(2),rad,th1,th12);
38:      if nargout > 1
39:        nplot=max(10,round(abs(th12*20/pi)));
40:        theta=th1+(0:nplot-1)'* ...
41:              (th12/(nplot-1));
42:        x=r(1)+rad*cos(theta);
43:        y=r(2)+rad*sin(theta);
44:        xb=[xb;x]; yb=[yb;y];
45:      end
46:    end
47: end
48: aprops(2)=aprops(2)/aprops(1);
49: aprops(3)=aprops(3)/aprops(1);
```

5.5.1.5   Function arcprop

```
1: function arcprp=arcprop(x0,y0,r0,t1,t12)
2: %
3: % arcprp = arcprop(x0,y0,r0,t1,t12)
4: % ~~~~~~~~~~~~~~~~~~~~~~~~~~~~~~~~~~~~
5: % This function computes area properties
6: % of a circular arc.
7: %
8: % (x0,y0)        - center coordinates
9: % r0             - arc radius
10: % t1            - polar angle of the
11: %                   starting point
12: % t12           - angle subtended at the
13: %                   center (positive direction
14: %                   is counterclockwise)
15: % arcprp        -[area;axbar;aybar;axx;axy;ayy]
16: % P(n,m)        - integral of
17: %                 (x^n y^m (x y'(t)-y x'(t)) dt;
18: %                 [t=t1 => t=t1+t12] )/(n+m+2)
19: % area          - P(0,0)
```

```
20: % [axbar,aybar] - [ P(1,0),P(0,1) ]
21: % [axx,axy,ayy] - [ P(2,0),P(1,1),P(0,2) ]
22: %
23: % User m functions called:  none
24: %---------------------------------------------------
25:
26: t2=t1+t12; s1=sin(t1); c1=cos(t1);
27: s2=sin(t2); c2=cos(t2);
28: fs=c1-c2; fc=s2-s1;
29: fss=.5*(t12-s2*c2+s1*c1); fcc=t12-fss;
30: fcs=.5*(s2+s1)*(s2-s1); fcss=(s2^3-s1^3)/3;
31: fccs=(c1^3-c2^3)/3; fccc=fc-fcss;
32: fsss=fs-fccs;
33: aa=x0*fc+y0*fs+r0*t12; area=aa*r0/2;
34: e1=(x0*fcc+y0*fcs+r0*fc)*r0;
35: e2=(x0*fcs+y0*fss+r0*fs)*r0;
36: axbar=(x0*aa+e1)*r0/3; aybar=(y0*aa+e2)*r0/3;
37: r2=r0/4; r4=(r0^3)/4;
38: axx=r2*x0*(x0*aa+2*e1)+ ...
39:     r4*(x0*fccc+y0*fccs+r0*fcc);
40: axy=r2*(x0*y0*aa+e1*y0+e2*x0)+ ...
41:     r4*(x0*fccs+y0*fcss+r0*fcs);
42: ayy=r2*y0*(y0*aa+2*e2)+ ...
43:     r4*(x0*fcss+y0*fsss+r0*fss);
44: arcprp=[area;axbar;aybar;axx;axy;ayy];
```

5.5.1.6 Function lineprop

```
 1: function [Lineprp]=lineprop(x,y)
 2: %
 3: % [Lineprp]=lineprop(x,y)
 4: % ~~~~~~~~~~~~~~~~~~~~~~~~
 5: % This function computes the area property
 6: % contributions associated with a polyline.
 7: %
 8: % x,y     - vectors containing data
 9: %             coordinates
10: % Lineprp - vector of geometrical properties
11: %             the components of which are
12: %             [area; axbar; aybar; axx; axy; ayy]
13: %
14: % User m functions called:  none
15: %---------------------------------------------------
```

```
16:
17: n=length(x); x=x(:)'; y=y(:)'; r=[x;y];
18: if n==2 % Case for a single line element
19:    area=det(r)/2;
20:    r1=r(:,1); r2=r(:,2); rs=r1+r2;
21:    arbar=(r1+r2)*area/3;
22:    arr=(r1*r1'+rs*rs'+r2*r2')*area;
23: else    % Case for a sequence of line segments
24:    J=[1:n-1]; J1=J+1;
25:    s=x(J).*y(J1)-y(J).*x(J1);
26:    area=sum(s)/2;
27:    arbar=(r(:,J)+r(:,J1)).*s([1 1],:);
28:    arbar=sum(arbar')/6;
29:    arr=zeros(2,2); rj=r(:,1); rrj=rj*rj';
30:    for j=1:n-1
31:       rj1=r(:,j+1); rrj1=rj1*rj1'; t=rj*rj1';
32:       arr=arr+(rrj+rrj1+(t+t')/2)*s(j);
33:       rj=rj1; rrj=rrj1;
34:    end
35: end
36: Lineprp=[area; arbar(:); ...
37: [arr(1,1); arr(1,2); arr(2,2)]/12];
```

5.5.1.7 Function pltundst

```
1:  function pltundst(x,y,grphnam)
2: %
3: % pltundst(x,y,grphnam)
4: % ~~~~~~~~~~~~~~~~~~~~~
5: % This function creates an undistorted
6: % plot of the curve defined by x,y.
7: %
8: % x,y     - data vectors defining the curve
9: % grphnam - character string defining the
10: %           graph title
11: %
12: % User m functions called:  none
13: %-----------------------------------------------
14:
15: xmin=min(x); xmax=max(x);
16: xwid=xmax-xmin; xc=(xmax+xmin)/2;
17: ymin=min(y); ymax=max(y);
18: ywid=ymax-ymin; yc=(ymax+ymin)/2;
```

```
19: w=1.2*max(xwid,ywid)/2;
20: scal=[xc-w; xc+w; yc-w; yc+w];
21: % plot(x,y,'w'); axis(scal); axis('square')
22: plot(x,y,'w',[xc-w,xc+w],[yc-w,yc+w],'i')
23: axis('equal')
24: xlabel('x axis'); ylabel('y axis');
25: if nargin > 2 , title(grphnam); end
26: axis('normal')
```

5.5.1.8   Function shftprop

```
 1: function AP = shftprop(ap,xctr,yctr,theta);
 2: %
 3: % AP = shftprop(ap,xctr,yctr,theta)
 4: % ~~~~~~~~~~~~~~~~~~~~~~~~~~~~~~~~~~~~
 5: % This function computes area properties for a
 6: % set of axes centered at (xctr,yctr) and
 7: % rotated counterclockwise through an angle
 8: % theta relative to the original axes.
 9: %
10: % ap - [ area; xbar; ybar; axx; axy; ayy ]
11: % AP - Transformed property vector for the
12: %       new axes
13: %
14: % User m functions called:  none
15: %-----------------------------------------------
16:
17: AP=ap(:); a=ap(1);
18: x=ap(2); y=ap(3); i=sqrt(-1);
19: if nargin==4
20:   X=x-xctr; Y=y-yctr; AP(2)=X; AP(3)=Y;
21:   AP(4)=AP(4)+a*(X*X-x*x);
22:   AP(5)=AP(5)+a*(X*Y-x*y);
23:   AP(6)=AP(6)+a*(Y*Y-y*y);
24: end
25: z=(AP(2)+i*AP(3))*exp(-i*theta);
26: AP(2)=real(z); AP(3)=imag(z);
27: p1=(AP(4)+AP(6))/2;
28: p2=(AP(4)-AP(6))/2+i*AP(5);
29: p2=p2*exp(-2*i*theta); AP(4)=real(p1+p2);
30: AP(5)=imag(p2); AP(6)=real(p1-p2);
```

5.5.2   PROGRAM TO ANALYZE SPLINE INTERPOLATED BOUNDARIES

The second program produces properties for a spline interpolated geometry. Function **makcrcsq** has been written to generate spline data suitable to approximate the exact geometry discussed above. Function **spcurv2d** is used to graph the spline interpolated geometry for both cases, with and without use of corner points. Results appear in Figures 5.5 and 5.6. The spline geometry in Figure 5.5 is used to reproduce Figure 5.2. The segments with corners at each end give perfect straight lines. The spline curve in Figure 5.6, which was obtained without use of corner points, is clearly unsatisfactory. Function **splaprop** computes geometrical properties for the spline interpolated boundary. Functions **spc** and **spltrp** are called to evaluate the necessary integrands. The numerical integration is performed using function **cbpwf6** and produces composite base points and weight factors in accord with the six point Gauss integration formula.

A comparison between the exact results produced by function **areaprop** and the spline results produced by **splaprop** is quite interesting. Table 5.2 indicates good agreement between both methods. The error from the spline method is entirely attributable to imperfect representation of the circular arcs. That error can be reduced by employing additional interpolation points to refine the arc definitions.

The programs developed in this chapter are adequate to analyze a variety of practical geometries. Furthermore, the results clearly show the power of cubic spline interpolation to describe general boundary shapes.

Property	Exact Results	Spline Results	Percent Error
area	16.7124	16.6899	0.1343
xbar	3.0000	3.0000	0.0000
ybar	4.1251	4.1221	0.0733
Axx	176.3020	175.9907	0.1766
Axy	206.8230	206.3937	0.2076
Ayy	359.5365	358.6261	0.2532

TABLE 5.2. Comparison of Area Properties

MATLAB EXAMPLE

5.5.2.1   Script File crnrtest

```
 1: % Example: crnrtest
 2: % ~~~~~~~~~~~~~~~~~
 3: % MATLAB example showing effect of corner
 4: % points on spline curve approximation of
 5: % a general geometry.
 6: %
 7: % User m functions required:
 8: %    makcrcsq, spcurv2d, splaprop, cbpwf6,
 9: %    spc, pltundst, removfil, spltrp, genprint
10: %-------------------------------------------------
11:
12: fprintf('\nAREA PROPERTIES OF A SPLINE ')
13: fprintf('INTERPOLATED GEOMETRY\n\n')
14: [xd,yd,icrnr]=makcrcsq;
15: [xxc,yyc]=spcurv2d(xd,yd,8,icrnr);
16: [xxnc,yync]=spcurv2d(xd,yd,8);
17:
18: pltundst(xxnc,yync, ...
19:          'PLOT WITHOUT CORNER POINTS')
20: hold on; axis('square'); plot(xd,yd,'ow');
21: disp('Press RETURN to continue'); pause
22: genprint('nocorner.ps'); clf; drawnow
23:
24: pltundst(xxc,yyc,'PLOT USING CORNER POINTS');
25: hold on; axis('square'); plot(xd,yd,'ow');
26: fprintf('\nPress RETURN to continue\n'); pause
27: genprint('crnrtst');
28: aprop=splaprop(xd,yd,icrnr);
29: fprintf('\n    area        xbar        ybar')
30: fprintf('         axx        axy        ayy\n')
31: disp(aprop(:)')
```

5.5.2.2   Function splaprop

```
 1: function aprop=splaprop(xd,yd,ic)
 2: %
 3: % aprop=splaprop(xd,yd,ic)
 4: % ~~~~~~~~~~~~~~~~~~~~~~~~~~~
```

```
 5: % This function computes geometrical properties
 6: % of an area bounded by a spline curve passed
 7: % through data points xd, yd. Vector aprop
 8: % contains the following quantities
 9: %
10: %    aprop - [ area; xbar; ybar; axx; axy; ayy]
11: %
12: % where (xbar,ybar) are centroidal coordinates,
13: % and
14: %
15: %    axx - integral of x*x*d(area)
16: %    axy - integral of x*y*d(area)
17: %    ayy - integral of y*y*d(area)
18: %
19: % User m functions called:  cbpwf6, spc, spltrp
20: %-------------------------------------------------
21:
22: n=length(xd); xd=xd(:); yd=yd(:);
23: if nargin==2, ic=[]; end
24: xp=[xd(2)-xd(n-1)]/2;
25: yp=[yd(2)-yd(n-1)]/2; id=[1;1];
26: s=(1:n)'; [bp,wf]=cbpwf6(s(1),s(n),n-1);
27: matx=spc(s,xd,id,xp*id,ic);
28: maty=spc(s,yd,id,yp*id,ic);
29: x=spltrp(bp,matx,0); xp=spltrp(bp,matx,1);
30: y=spltrp(bp,maty,0); yp=spltrp(bp,maty,1);
31: mat=zeros(length(x),6);
32: mat(:,1)=x.*yp-y.*xp; mat(:,2)=mat(:,1).*x;
33: mat(:,3)=mat(:,1).*y; mat(:,4)=mat(:,2).*x;
34: mat(:,5)=mat(:,2).*y; mat(:,6)=mat(:,3).*y;
35: aprop=wf'*mat;
36: aprop=aprop(:)./[2; 3; 3; 4; 4; 4];
37: aprop(2)=aprop(2)/aprop(1);
38: aprop(3)=aprop(3)/aprop(1);
```

5.5.2.3 Function makcrcsq

```
1: function [x,y,icrnr]=makcrcsq
2: %
3: % [x,y,icrnr]=makcrcsq
4: % ~~~~~~~~~~~~~~~~~~~~~
5: % This function creates data for a geometry
6: % involving half of an annulus placed above a
```

```
 7: % square containing a square hole.
 8: %
 9: % x,y   - data points characterizing the data
10: % icrnr - index vector defining corner points
11: %
12: % User m functions called:  none
13: %-----------------------------------------------
14:
15: xshift=3.0; yshift=3.0;
16: a=2; b=1; narc=7; x0=0; y0=2*a-b;
17: xy=[ a,   a
18:      -a,  a
19:      -b,  b
20:       b,  b
21:       b, -b
22:      -b, -b
23:      -b,  b
24:      -a,  a
25:      -a, -a
26:       a, -a
27:       a,  a];
28: theta=linspace(0,pi,narc);
29: c=cos(theta); s=sin(theta);
30: c=c(:); s=s(:);
31: xy=[xy;[x0+a*c,y0+a*s]];
32: c=flipud(c); s=flipud(s);
33: xy=[xy;[x0+b*c,y0+b*s];[a,y0];[a,a]];
34: x=xy(:,1)+xshift; y=xy(:,2)+yshift;
35: icrnr=[(1:12)';11+narc;12+narc;  ...
36:         11+2*narc;12+2*narc;13+2*narc];
```

5.5.2.4   Function cbpwf6

```
 1: function [cbp,cwf]=cbpwf6(xlow,xhigh,mparts)
 2: %
 3: % [cbp,cwf]=cbpwf6(xlow,xhigh,mparts)
 4: % ~~~~~~~~~~~~~~~~~~~~~~~~~~~~~~~~~~~~~
 5: % This function computes base points, cbp,
 6: % and weight factors, cwf, in a composite
 7: % quadrature formula which integrates an
 8: % arbitrary function from xlow to xhigh by
 9: % dividing the interval of integration into
10: % mparts equal parts and integrating over
```

176

```
11: % each part using a Gauss formula requiring
12: % six function values.
13: %
14: % xlow, xhigh - integration limits
15: % mparts      - number of subintervals used
16: %                for composite integration
17: % cbp, cwf    - vectors of length 6*mparts
18: %                which contain the composite
19: %                base points and weight factors
20: %
21: % User m functions called:  none
22: %-------------------------------------------------
23:
24: wf=[ 1.71324492379170d-01; ...
25:      3.60761573048139d-01; ...
26:      4.67913934572691d-01];
27: wf=[wf;wf([3 2 1])];
28: bp=[-9.32469514203152d-01; ...
29:     -6.61209386466265d-01;...
30:     -2.38619186083197d-01];
31: bp=[bp;-bp([3 2 1])];
32:
33: d=(xhigh-xlow)/mparts; d1=d/2; nquad=6;
34: cbp=(d1*bp)*ones(1,mparts)+ ...
35:     (d*ones(nquad,1))*(1:mparts);
36: cbp=cbp(:)+(xlow-d1);
37: cwf=(d1*wf)*ones(1,mparts);
38: cwf=cwf(:);
```

6

Fourier Series and the FFT

6.1 Definitions and Computation of Fourier Coefficients

Trigonometric series are useful to represent periodic functions. A function defined for $-\infty < x < \infty$ has a period of 2π if $f(x+2\pi) = f(x)$ for all x. In most practical situations, such a function can be expressed as a complex Fourier series

$$f(x) = \sum_{j=-\infty}^{\infty} c_j e^{\imath j x} \qquad \imath = \sqrt{-1}$$

The numbers c_j, called complex Fourier coefficients, are computed by integration as

$$c_j = \frac{1}{2\pi} \int_0^{2\pi} f(x) e^{-\imath j x} dx$$

The Fourier series can also be rewritten using sines and cosines as

$$f(x) = c_0 + \sum_{j=1}^{\infty} (c_j + c_{-j}) \cos(jx) + \imath(c_j - c_{-j}) \sin(jx)$$

Denoting

$$a_j = c_j + c_{-j} \qquad b_j = \imath(c_j - c_{-j})$$

yields

$$f(x) = \frac{1}{2} a_0 + \sum_{j=1}^{\infty} a_j \cos(jx) + b_j \sin(jx)$$

which is called a Fourier sine-cosine expansion. This series is most appealing when $f(x)$ is real valued. For that case $c_{-j} = \bar{c}_j$ for all j, which implies that c_0 must be real and

$$a_j = 2 \, \mathbf{real}(c_j) \qquad b_j = -2 \, \mathbf{imag}(c_j) \qquad j > 0$$

Suppose we want a Fourier series expansion for a more general function $f(x)$ having period p instead of 2π. If we introduce a new function $g(x)$ defined by

$$g(x) = f(\frac{px}{2\pi})$$

then $g(x)$ has a period of 2π. Consequently, $g(x)$ can be represented as

$$g(x) = \sum_{j=-\infty}^{\infty} c_j e^{ijx}$$

From the fact that $f(x) = g(2\pi x/p)$ we deduce that

$$f(x) = \sum_{j=-\infty}^{\infty} c_j e^{2\pi ijx/p}$$

A need sometimes occurs to expand a function as a series of sine terms only, or as a series of cosine terms only. If the function is originally defined for $0 < x < \frac{p}{2}$, then making $f(x) = -f(p-x)$ for $\frac{p}{2} < x < p$ gives a series involving only sine terms. Similarly, if $f(x) = +f(p-x)$ for $\frac{p}{2} < x < p$, only cosine terms arise. Thus we get

$$f(x) = c_0 + \sum_{j=1}^{\infty}(c_j + c_{-j})\cos(2\pi jx/p) \qquad \text{if:} \quad f(x) = f(p-x)$$

or

$$f(x) = \sum_{j=1}^{\infty} i(c_j - c_{-j})\sin(2\pi jx/p) \qquad \text{if:} \quad f(x) = -f(p-x)$$

When the Fourier series of a function is approximated using a finite number of terms, the resulting approximating function may oscillate in regions where the actual function is discontinuous or changes rapidly. This undesirable behavior can be reduced by using a smoothing procedure described by Lanczos [50]. Use is made of Fourier series of a closely related function $\hat{f}(x)$ defined by a local averaging process according to

$$\hat{f}(x) = \frac{1}{\Delta}\int_{x-\frac{\Delta}{2}}^{x+\frac{\Delta}{2}} f(\zeta)d\zeta$$

where the averaging interval Δ should be a small fraction of the period p. Hence we write $\Delta = \alpha p$ with $\alpha < 1$. The functions $\hat{f}(x)$ and $f(x)$ are identical as $\alpha \to 0$. Even for $\alpha > 0$, these functions also match exactly at any point x where $f(x)$ varies linearly between $x - \frac{\Delta}{2}$ and $x + \frac{\Delta}{2}$. An important property of $\hat{f}(x)$ is that it agrees closely with $f(x)$ for small α but has a Fourier series which converges more rapidly than the series for $f(x)$. Furthermore, from its definition,

$$\hat{f}(x) = \sum_{j=-\infty}^{\infty} c_j \frac{1}{p\alpha} \int_{x-\frac{p\alpha}{2}}^{x+\frac{p\alpha}{2}} e^{2\pi \imath j x/p} \, dx$$

$$= \sum_{j=-\infty}^{\infty} \hat{c}_j e^{2\pi \imath j x/p}$$

where $\hat{c}_0 = c_0$ and $\hat{c}_j = c_j \sin(\pi j \alpha)/(\pi j \alpha)$ for $j \neq 0$. Evidently the Fourier coefficients of $\hat{f}(x)$ are easily obtainable from those of $f(x)$. When the series for $f(x)$ converges slowly, using the same number of terms in the series for $\hat{f}(x)$ often gives an approximation preferable to that provided by the series for $f(x)$. This process is called smoothing.

6.1.1 TRIGONOMETRIC INTERPOLATION AND THE FFT

Computing Fourier coefficients by numerical integration is very time consuming. Consequently, we are led to investigate alternative methods employing trigonometric polynomial interpolation through evenly spaced data. The resulting formulas are the basis of an important algorithm called the FFT (Fast Fourier Transform). Although the Fourier coefficients obtained by interpolation are approximate, these coefficients can be computed very rapidly when the number of sample points is an integer power of 2. We will discuss next the ideas behind trigonometric polynomial interpolation among evenly spaced data values.

Suppose we truncate the Fourier series and only use harmonics up to some order N. We assume $f(x)$ has period 2π so

$$f(x) = \sum_{j=-N}^{N} c_j e^{\imath j x}$$

This trigonometric polynomial satisfies $f(0) = f(2\pi)$ even though the original function might actually have a finite discontinuity at 0 and 2π. Consequently, we may choose to use, in place of $f(0)$, the limit as $\epsilon \to 0$ of $[f(\epsilon) + f(2\pi - \epsilon)]/2$.

It is well known that the functions e^{ijx} satisfy an orthogonality condition for integration over the interval 0 to 2π. They also satisfy an orthogonality condition regarding summation over equally spaced data. The latter condition is useful for deriving a discretized approximation of the integral formula for the exact Fourier coefficients. Let us choose data points

$$x_j = (\frac{2\pi}{2N})j = (\frac{\pi}{N})j \qquad 0 \le j \le (2N-1)$$

and write the simultaneous equations to make the trigonometric polynomial match the original function at the equally spaced data points. To shorten the notation we let

$$t = e^{i\pi/N} \qquad t^k = e^{ik\pi/N}$$

and write

$$f_k = \sum_{j=-N}^{N} c_j t^{kj}$$

Suppose we pick an arbitrary integer n in the range $-N < n < N$. Multiplying the last equation by t^{-kn} and summing from $k = 0$ to $2N - 1$ gives

$$\sum_{k=0}^{2N-1} f_k t^{-kn} = \sum_{k=0}^{2N-1} t^{-kn} \sum_{j=-N}^{N} c_j t^{kj}$$

Interchanging the summation order in the last equation yields

$$\sum_{k=0}^{2N-1} f_k t^{-kn} = \sum_{j=-N}^{N} c_j \sum_{k=0}^{2N-1} \zeta^k$$

when $x_i = e^{i(j-n)\pi/N}$. Summing the inner geometric series gives

$$\sum_{k=0}^{2N-1} \zeta^k = \begin{cases} \frac{1-\zeta^{2N}}{1-\zeta} & \text{for } \zeta \ne 1 \\ 2N & \text{for } \zeta = 1 \end{cases}$$

We find, for all k and n in the stated range, that

$$\zeta^{2N} = e^{i2\pi(k-n)} = 1$$

Therefore we get

$$\sum_{k=0}^{2N-1} f_k t^{-kn} = 2Nc_n \qquad \text{for} - N < n < N$$

In the cases where $n = \pm N$, the procedure just outlined only gives a relationship governing $c_N + c_{-N}$. Since the first and last terms cannot be computed uniquely, we customarily take N large enough to discard these last two terms and write simply

$$c_n = \frac{1}{2N} \sum_{k=0}^{2N-1} f_k t^{-kn} \qquad - N < n < N$$

This formula is the basis for fast algorithms (called FFT for Fast Fourier Transform) to compute approximate Fourier coefficients. The periodicity of the terms depending on various powers of $e^{i\pi/N}$ can be utilized to greatly reduce the number of trigonometric function evaluations. The case where N equals a power of 2 is especially attractive. The mathematical development is not provided here. However, the related theory was presented by Cooley and Tukey in 1965 [18] and has been expounded in many textbooks [45, 82]. The result is a remarkably concise algorithm which can be comprehended without studying the details of the mathematical derivation. For our present interests it is important to understand how to use MATLAB's intrinsic function for the FFT (**fft**).

Suppose a periodic function is evaluated at a number of equidistant points ranging over one period. It is preferable for computational speed that the number of sample points should equal an integer power of two ($n = 2^m$). Let the function values for argument vector

```
x=p/n*(0:n-1)
```

be an array f denoted by

$$f \iff [f_1, f_2, \cdots, f_n]$$

The function evaluation $\mathbf{fft}(f)$ produces an array of complex Fourier coefficients multiplied by n and arranged in a peculiar fashion. Let us illustrate this result for $n = 8$. If $f = [f_1, f_2, \cdots, f_8]$, then $\mathbf{fft}(f)/8$ produces $c = [c_0, c_1, c_2, c_3, c_*, c_{-3}, c_{-2}, c_{-1}]$. The term denoted by c_* actually turns out to equal $c_4 + c_{-4}$, so it would not be used in subsequent calculations. We generalize this procedure for arbitrary n as follows. Let $N = n/2 - 1$. In the transformed array, elements with indices of $1, \cdots, N+1$ correspond to $c_0, \cdots, c_N$ and elements with indices of $n, n-1, n-2, \cdots, N+3$ correspond to $c_{-1}, c_{-2}, c_{-3}, \cdots, c_{-N}$. It is also useful to remember that a real valued function has $c_{-n} = \mathbf{conj}(c_n)$. To fix our ideas about how to evaluate a Fourier series, suppose we want to sum an approximation involving harmonics from order zero to order $(nsum - 1)$. We are dealing with a real valued function defined by **func** with a real argument vector x. The following code expands **func** and sums the series for argument x using $nsum$ terms.

```
function fouval=fftaprox(func,period,nfft,nsum,x)
fc=feval(func,period/nfft*(0:nfft-1));
fc=fft(fc)/nfft; fc(1)=fc(1)/2;
w=2*pi/period*(0:nsum-1);
fouval=2*real(exp(i*x(:)*w)*fc(:));
```

6.2   Some Applications

Applications of Fourier series arise in numerous practical situations such as structural dynamics, signal analysis, solution of boundary value problems, and image processing. Three examples are given below which illustrate use of the FFT. The first example calculates Bessel functions and the second problem studies forced dynamic response of a lumped mass system. The final example presents a program for constructing Fourier expansions and displaying graphical results for linearly interpolated or analytically defined functions.

6.2.1   USING THE FFT TO COMPUTE INTEGER ORDER BESSEL FUNCTIONS

The FFT provides an efficient way to compute integer order Bessel functions $J_n(z)$ which are important in various physical applications [99]. Function $J_n(z)$ can be obtained as the complex Fourier coefficient of $e^{in\theta}$ in the generating function described by

$$e^{iz\sin(\theta)} = \sum_{n=-\infty}^{\infty} J_n(z)e^{in\theta}$$

Orthogonality conditions imply

$$J_n(z) = \frac{1}{2\pi} \int_0^{2\pi} e^{i(z\sin(\theta)-n\theta)} \, d\theta$$

The Fourier coefficients represented by $J_n(Z)$ can be computed approximately with the FFT. The infinite series converges very rapidly because the function it represents has continuous derivatives of all finite orders. Of course, $e^{iz\sin(\theta)}$ is highly oscillatory for large $|z|$, thereby requiring a large number of sample points in the FFT to obtain accurate results. For $n < 30$ and $|z| < 30$, a 128-point transform is adequate to give about ten digit accuracy for values of $J_n(z)$. The following code implements the above ideas and plots a surface showing how J_n changes in terms of n and z.

Surface Plot For Jn(x)

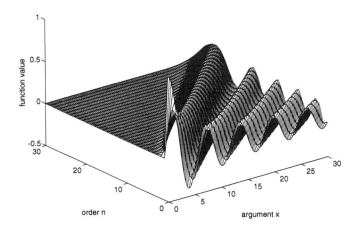

FIGURE 6.1. Surface Plot for $J_n(X)$

MATLAB Example

6.2.1.1   Script File plotjrun

```
1: % Example: plotjrun
2: % ~~~~~~~~~~~~~~~~~
3: % This program computes integer order Bessel
4: % functions of the first kind by use of the
5: % FFT. The algorithm is often faster than
6: % function BESSELN provided in MATLAB.
7: %
8: % User m functions required:
9: %     jnft
10: %-------------------------------------------------
11:
12: x=0:.3:30; n=0:30; [J,tcp]=jnft(n,x);
13: mesh(x,n,J'); title('Surface Plot For Jn(x)')
14: ylabel('order n'); xlabel('argument x');
15: zlabel('function value')
16: view(3); axis([0 30 0 30 -.5 1]);
```

6.2.1.2   Function jnft

```
1: function [J,tcp]=jnft(n,z,nft)
2: %
3: % [J,tcp]=jnft(n,z,nft)
4: % ~~~~~~~~~~~~~~~~~~~~~
5: % Integer order Bessel functions of the
6: % first kind computed by use of the Fast
7: % Fourier Transform (FFT).
8: %
9: % n   - integer vector defining the function
10: %        orders
11: % z   - a vector of values defining the
12: %        arguments
13: % nft - number of function evaluations used
14: %        in the FFT calculation. This value
15: %        should be an integer power of 2 and
16: %        should exceed twice the largest
17: %        component of n. When nft is omitted
18: %        from the argument list, then a value
19: %        equal to 128 is used. More accurate
```

```
20: %          values of J are computed as nft is
21: %          increased. For max(n) < 30 and
22: %          max(z) < 30, nft=128 gives about
23: %          ten digit accuracy.
24: % J    - a matrix of values for the integer
25: %          order Bessel function of the first
26: %          kind. Row position matches orders
27: %          defined by n, and column position
28: %          corresponds to arguments defined by
29: %          components of z.
30: % tcp - computer time required to make the
31: %          calculation
32: %
33: % User m functions called:  none.
34: %---------------------------------------------
35:
36: tcp=cputime;
37: if nargin<3, nft=128; end;
38: J=exp(sin((0:nft-1)'* ...
39:   (2*pi/nft))*(i*z(:).'))/nft;
40: J=fft(J); J=J(1+n,:).';
41: if sum(abs(imag(z)))<max(abs(z))/1e10
42:   J=real(J);
43: end
44: tcp=cputime-tcp;
```

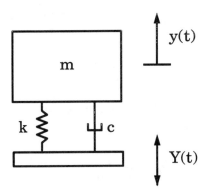

FIGURE 6.2. Mass System

6.2.2   DYNAMIC RESPONSE OF A MASS ON AN OSCILLATING FOUNDATION

Fourier series are often used to describe time dependent phenomena such as earthquake ground motion. Understanding the effects of foundation motions on an elastic structure is important in design. The model in Figure 6.2 embodies rudimentary aspects of this type of system and consists of a concentrated mass connected by a spring and viscous damper to a base which oscillates with known motion $Y(t)$. The system is assumed to have arbitrary initial conditions $y(0) = y_0$ and $\dot{y}(0) = v_0$ when the base starts moving. The resulting displacement and acceleration of the mass are to be computed.

We assume that $Y(t)$ can be represented well over some time interval p by a Fourier series of the form

$$Y(t) = \sum_{n=-\infty}^{\infty} c_n e^{\imath \omega_n t} \qquad \omega_n = \frac{2n\pi}{p}$$

where $c_{-n} = \mathbf{conj}(c_n)$ because Y is real valued. The differential equation governing this problem is

$$m\ddot{y} + c\dot{y} + ky = kY(t) + c\dot{Y}(t) = F(t)$$

where the forcing function can be expressed as

$$F(t) = \sum_{n=-\infty}^{\infty} c_n[k + \imath c\omega_n]e^{\imath \omega_n t}$$

$$= kc_0 + 2\,\mathbf{real}(\sum_{n=1}^{\infty} f_n e^{\imath \omega_n t})$$

and

$$f_n = c_n(k + \imath c \omega_n)$$

The corresponding steady-state solution of the differential equation is representable as

$$y_s(t) = \sum_{n=-\infty}^{\infty} y_n e^{\imath \omega_n t}$$

where $y_{-n} = \mathbf{conj}(y_n)$ because $y_s(t)$ is real valued. Substituting the series solution into the differential equation and comparing coefficients of $e^{\imath \omega_n t}$ on both sides leads to

$$y_n = \frac{c_n(k + \imath c \omega_n)}{k - m\omega_n^2 + \imath c \omega_n}$$

These coefficients satisfy $y_{-n} = \mathbf{conj}(y_n)$, so the displacement, velocity, and acceleration corresponding to the steady-state (also called particular) solution are

$$y_s(t) \;\; = \;\; c_0 + 2\,\mathbf{real}(\sum_{n=1}^{\infty} y_n e^{\imath \omega_n t})$$

$$\dot{y}_s(t) \;\; = \;\; 2\,\mathbf{real}(\sum_{n=1}^{\infty} \imath \omega_n y_n e^{\imath \omega_n t})$$

$$\ddot{y}_s(t) \;\; = \;\; -2\,\mathbf{real}(\sum_{n=1}^{\infty} \omega_n^2 y_n e^{\imath \omega_n t})$$

The initial conditions satisfied by y_s are

$$y_s(0) \;\; = \;\; c_0 + 2\,\mathbf{real}(\sum_{n=1}^{\infty} y_n)$$

$$\dot{y}_s(0) \;\; = \;\; 2\,\mathbf{real}(\sum_{n=1}^{\infty} \imath \omega_n y_n)$$

Because these values usually will not match the desired initial conditions, the total solution consists of $y_s(t)$ plus another function $y_h(t)$ satisfying the homogeneous differential equation

$$m\ddot{y}_h + c\dot{y}_h + ky_h = 0$$

The solution is

$$y_h = g_1 e^{s_1 t} + g_2 e^{s_2 t}$$

where s_1 and s_2 are roots satisfying

$$ms^2 + cs + k = 0$$

The roots are

$$s_1 = \frac{-c + \sqrt{c^2 - 4mk}}{2m} \qquad s_2 = \frac{-c - \sqrt{c^2 - 4mk}}{2m}$$

Since the total solution is

$$y(t) = y_s(t) + y_h(t)$$

the constants g_1 and g_2 are obtained by solving the two simultaneous equations

$$g_1 + g_2 = y(0) - y_s(0)$$

$$s_1 g_1 + s_2 g_2 = \dot{y}(0) - \dot{y}_s(0)$$

The roots s_1 and s_2 are equal when $c = 2\sqrt{mk}$. Then the homogeneous solution assumes an alternate form given by $(g_1 + g_2 t)e^{st}$ with $s = -c/(2m)$. In this special case we find that

$$g_1 = y(0) - y_s(0) \qquad g_2 = \dot{y}(0) - \dot{y}_s(0) - sg_1$$

It should be noted that even though roots s_1 and s_2 will often be complex numbers, this causes no difficulty since MATLAB handles the complex arithmetic automatically (just as it does when the FFT transforms real function values into complex Fourier coefficients).

The harmonic response solution works satisfactorily for a general forcing function as long as the damping coefficient c is nonzero. A special situation can occur when $c = 0$, because the forcing function may resonate with the natural frequency of the undamped system. If c is zero, and for some n we have $\sqrt{k/m} = 2\pi n/p$, a condition of harmonic resonance is produced and a value of zero in the denominator occurs when the corresponding y_n is computed. What actually happens in the undamped resonant case is the particular solution grows like $[te^{\imath \omega_n t}]$, quickly becoming large. Even when c is small and $\sqrt{k/m} \approx 2\pi n/p$, undesirably large values of y_n can result. Readers interested in the important phenomenon of resonance can find more detail in Meirovitch [58].

This example concludes by using a base motion resembling an actual earthquake excitation. Seismograph output employing about 2700 points recorded during the Imperial Valley, California, earthquake of 1940 provided the displacement history for Figure 6.3. The period used to describe the motion is 53.8 seconds. A program was written to analyze system response due to a simulated earthquake base excitation. The following program modules are used:

runimpv	sets data values and generates graphical results
fouaprox	generates Fourier series approximations for a general function
imptp	piecewise linear function approximating the Imperial Valley earthquake data
shkbftss	computes steady-state displacement and acceleration for a spring-mass-dashpot system subjected to base motion expandable in a Fourier series
hsmck	computes the homogeneous solution for the spring-mass-dashpot system subjected to general initial conditions

Numerical results were obtained for a system having a natural period close to one second ($2\pi/6 \approx 1.047$) and a damping factor of 5 percent. The function **imptp** was employed as an alternative to the actual seismograph data to provide a concisely expressible function which still embodies characteristics of a realistic base motion. Figure 6.4 shows a plot of function **imptp** along with its approximation by a twenty-term Fourier series. The series representation is surprisingly good considering the fact that such a

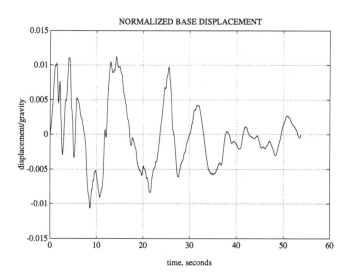

FIGURE 6.3. Normalized Base Displacement

small number of terms is used. The use of two-hundred terms gives an approximation which graphically does not deviate perceptibly from the actual function. Results showing how rapidly the Fourier coefficients diminish in magnitude with increasing order appear in Figure 6.5. The dynamical analysis produced displacement and acceleration values for the mass. Figure 6.6 shows both the total displacement as well as the displacement contributed from the homogeneous solution alone. Evidently, the steady-state harmonic response function captures well most of the motion and the homogeneous part could probably be neglected without serious error. Figure 6.7 also shows the total acceleration of the mass which is, of course, proportional to the resultant force on the mass due to the base motion.

Before proceeding to the next example, the reader should be sure to appreciate the following important fact. Once a truncated Fourier series expansion of the forcing function using some appropriate number of terms is chosen, the truncated series defines an input function for which the response is computed exactly. If the user takes enough terms in the truncated series so that he/she is well satisfied with the function it approximates, then the com-

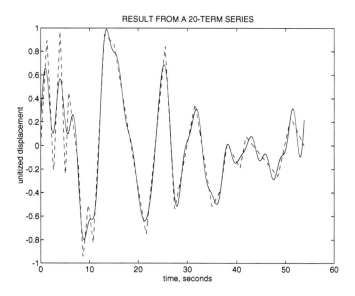

FIGURE 6.4. 20-Term Series

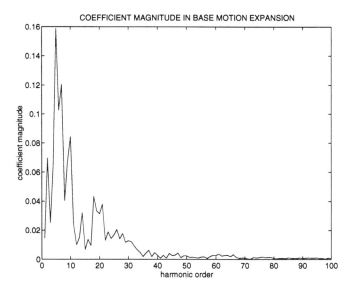

FIGURE 6.5. Coefficient Magnitude in Base Motion Expansion

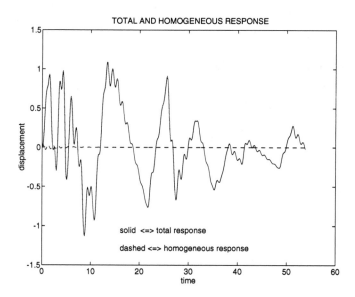

FIGURE 6.6. Total and Homogeneous Response

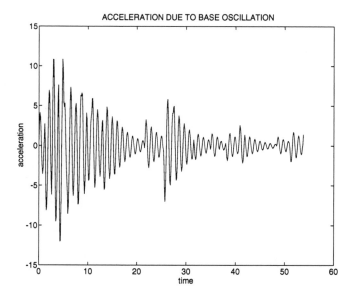

FIGURE 6.7. Acceleration Due to Base Oscillation

puted response value for $y(t)$ will also be acceptable. This situation is distinctly different from the more complicated type of approximations occurring when finite difference or finite element methods produce discrete approximations for continuous field problems. Understanding the effects of grid size discretization error is more complex than understanding the effects of series truncation in the example given here.

MATLAB EXAMPLE

6.2.2.1 Script file runimpv

```
 1: % Example:  runimpv
 2: % ~~~~~~~~~~~~~~~~~~
 3: % This is a driver program for the
 4: % earthquake example.
 5: %
 6: % User m functions required:
 7: %     fouaprox, removfil, imptp, hsmck,
 8: %     shkbftss, fouseris, lintrp, genprint
 9: %-------------------------------------------------
10:
11: % Make the undamped period about one
12: % second long
13: m=1; k=36;
14:
15: % Use damping equal to 5 percent of critical
16: c=.05*(2*sqrt(m*k));
17:
18: % Choose a period equal to length of
19: % Imperial Valley earthquake data
20: prd=53.8;
21:
22: nft=1024; tmin=0; tmax=prd;
23: ntimes=501; nsum=200;
24: tplt=linspace(0,prd,ntimes);
25: y20trm=fouaprox('imptp',prd,tplt,20);
26: plot(tplt,y20trm,'w',tplt,imptp(tplt),'w--');
27: xlabel('time, seconds')
28: ylabel('unitized displacement');
29: title('RESULT FROM A 20-TERM SERIES')
30: disp('Press RETURN to continue'); pause;
31: genprint('20trmplt');
32:
33: % Show how magnitudes of Fourier coefficients
34: % decrease with increasing harmonic order
35:
36: fcof=fft(imptp((0:1023)/1024,1))/1024;
37: plot(abs(fcof(1:100)),'w');
38: xlabel('harmonic order')
39: ylabel('coefficient magnitude');
40: title(['COEFFICIENT MAGNITUDE IN BASE ' ...
```

```
41:        'MOTION EXPANSION']);
42: disp('Press RETURN to continue'); pause;
43: genprint('coefsize');
44:
45: % Compute forced response
46: [t,ys,ys0,vs0,as]= ...
47:   shkbftss(m,c,k,'imptp',prd,nft,nsum, ...
48:             tmin,tmax,ntimes);
49:
50: % Compute homogeneous solution
51: [t,yh,ah]= ...
52:   hsmck(m,c,k,-ys0,-vs0,tmin,tmax,ntimes);
53:
54: % Obtain the combined solution
55: y=ys(:)+yh(:); a=as(:)+ah(:);
56: plot(t,y,'w',t,yh,'w--');
57: xlabel('time'); ylabel('displacement')
58: h=gca;
59: r(1:2)=get(h,'XLim'); r(3:4)=get(h,'YLim');
60: dx=r(1)+(r(2)-r(1))/10;
61: dy1=r(3)+(r(4)-r(3))*.15;
62: dy2=r(3)+(r(4)-r(3))*.075;
63: text(dx,dy1,'solid  <=> total response')
64: text(dx,dy2,'dashed <=> homogeneous response')
65: title('TOTAL AND HOMOGENEOUS RESPONSE');
66: disp('Press RETURN to continue'); pause;
67: genprint('displac');
68:
69: plot(t,a,'w');
70: xlabel('time'); ylabel('acceleration')
71: title('ACCELERATION DUE TO BASE OSCILLATION')
72: genprint('accel');
```

6.2.2.2 Function fouaprox

```
1: function y=fouaprox(func,per,t,nsum,nft)
2: %
3: % y=fouaprox(func,per,t,nsum,nft)
4: %   ~~~~~~~~~~~~~~~~~~~~~~~~~~~~~~~~~
5: % Approximation of a function by a Fourier
6: % series.
7: %
8: % func   - function being expanded
```

```
 9: % per    - period of the function
10: % t      - vector of times at which the series
11: %          is to be evaluated
12: % nsum   - number of terms summed in the series
13: % nft    - number of function values used to
14: %          compute Fourier coefficients. This
15: %          should be an integer power of 2.
16: %          The default is 1024
17: %
18: % User m functions called:  none.
19: %-------------------------------------------------
20:
21: if nargin<5, nft=1024; end;
22: nsum=min(nsum,fix(nft/2));
23: c=fft(feval(func,per/nft*(0:nft-1)))/nft;
24: c(1)=c(1)/2;
25: c=c(:); c=c(1:nsum);
26: w=2*pi/per*(0:nsum-1);
27: y=2*real(exp(i*t(:)*w)*c);
```

6.2.2.3   Function imptp

```
 1: function ybase=imptp(t,period)
 2: %
 3: % ybase=imptp(t,period)
 4: % ~~~~~~~~~~~~~~~~~~~~~
 5: % This function defines a piecewise linear
 6: % function resembling the ground motion of
 7: % the earthquake which occurred in 1940 in
 8: % the Imperial Valley of California. The
 9: % maximum amplitude of base motion is
10: % normalized to equal unity.
11: %
12: % period - period of the motion
13: %          (optional argument)
14: % t      - vector of times between
15: %          tmin and tmax
16: % ybase  - piecewise linearly interpolated
17: %          base motion
18: %
19: % User m functions called:  lintrp
20: %-------------------------------------------------
21:
```

```
22: tft=[ ...
23:    0.00     1.26     2.64     4.01     5.10 ...
24:    5.79     7.74;    8.65     9.74    10.77 ...
25:   13.06    15.07    21.60    25.49;   27.38 ...
26:   31.56    34.94    36.66    38.03    40.67 ...
27:   41.87;   48.40    51.04    53.80     0   ...
28:    0        0        0  ]';
29: yft=[ ...
30:    0        0.92    -0.25     1.00    -0.29 ...
31:    0.46    -0.16;   -0.97    -0.49    -0.83 ...
32:    0.95     0.86    -0.76     0.85;   -0.55 ...
33:    0.36    -0.52    -0.38     0.02    -0.19 ...
34:    0.08;   -0.26     0.24     0.00     0   ...
35:    0        0        0  ]';
36: tft=tft(:); yft=yft(:);
37: tft=tft(1:24); yft=yft(1:24);
38: if nargin == 2
39:   tft=tft*period/max(tft);
40: end
41: ybase=lintrp(t,tft,yft);
```

6.2.2.4   Function shkbftss

```
1: function [t,ys,ys0,vs0,as]=...
2:     shkbftss(m,c,k,ybase,prd,nft,nsum, ...
3:                  tmin,tmax,ntimes)
4: %
5: % [t,ys,ys0,vs0,as]=...
6: %    shkbftss(m,c,k,ybase,prd,nft,nsum, ...
7: %                 tmin,tmax,ntimes)
8: % ~~~~~~~~~~~~~~~~~~~~~~~~~~~~~~~~~~~~~~~~~~~~
9: % This function determines the steady state
10: % solution of the scalar differential equation
11: %
12: %    m*y''(t) + c*y'(t) + k*y(t) =
13: %                    k*ybase(t) + c*ybase'(t)
14: %
15: % where ybase is a function of period prd
16: % which is expandable in a Fourier series
17: %
18: % m,c,k      - Mass, damping coefficient, and
19: %                spring stiffness
20: % ybase      - Function or vector of
```

```
21: %               displacements equally spaced in
22: %               time which describes the base
23: %               motion over a period
24: % prd        - Period used to expand xbase in a
25: %               Fourier series
26: % nft        - The number of components used
27: %               in the FFT (should be a power
28: %               of two). If nft is input as
29: %               zero, then ybase must be a
30: %               vector and nft is set to
31: %               length(ybase)
32: % nsum       - The number of terms to be used
33: %               to sum the Fourier series
34: %               expansion of ybase. This should
35: %               not exceed nft/2.
36: % tmin,tmax  - The minimum and maximum times
37: %               for which the solution is to
38: %               be computed
39: % t          - A vector of times at which
40: %               the solution is computed
41: % ys         - Vector of steady state solution
42: %               values
43: % ys0,vs0    - Position and velocity at t=0
44: % as         - Acceleration ys''(t), if this
45: %               quantity is required
46: %
47: % User m functions called:  none.
48: %----------------------------------------------
49:
50: if nft==0
51:    nft=length(ybase); ybft=ybase(:)
52: else
53:    tbft=prd/nft*(0:nft-1);
54:    ybft=fft(feval(ybase,tbft))/nft;
55:    ybft=ybft(:);
56: end
57: nsum=min(nsum,fix(nft/2)); ybft=ybft(1:nsum);
58: w=2*pi/prd*(0:nsum-1);
59: t=tmin+(tmax-tmin)/(ntimes-1)*(0:ntimes-1)';
60: etw=exp(i*t*w); w=w(:);
61: ysft=ybft.*(k+i*c*w)./(k+w.*(i*c-m*w));
62: ysft(1)=ysft(1)/2;
63: ys=2*real(etw*ysft); ys0=2*real(sum(ysft));
64: vs0=2*real(sum(i*w.*ysft));
65: if nargout > 4
```

```
66:   ysft=-ysft.*w.^2;
67:   as=2*real(etw*ysft);
68: end
```

6.2.2.5 Function hsmck

```
 1: function [t,yh,ah]= ...
 2:           hsmck(m,c,k,y0,v0,tmin,tmax,ntimes)
 3: %
 4: % [t,yh,ah]=hsmck(m,c,k,y0,v0,tmin,tmax,ntimes)
 5: % ~~~~~~~~~~~~~~~~~~~~~~~~~~~~~~~~~~~~~~~~~~~~~~~
 6: % Solution of
 7: %     m*yh''(t) + c*yh'(t) + k*yh(t) = 0
 8: % subject to initial conditions of
 9: %     yh(0) = y0 and yh'(0) = v0
10: %
11: % m,c,k      -  mass, damping and spring
12: %               constants
13: % y0,v0      -  initial position and velocity
14: % tmin,tmax  -  minimum and maximum times
15: % ntimes     -  number of times to evaluate
16: %               solution
17: % t          -  vector of times
18: % yh         -  displacements for the
19: %               homogeneous solution
20: % ah         -  accelerations for the
21: %               homogeneous solution
22: %
23: % User m functions called:  none.
24: %---------------------------------------------
25:
26: t=tmin+(tmax-tmin)/(ntimes-1)*(0:ntimes-1);
27: r=sqrt(c*c-4*m*k);
28: if r~=0
29:   s1=(-c+r)/(2*m); s2=(-c-r)/(2*m);
30:   g=[1,1;s1,s2]\[y0;v0];
31:   yh=real(g(1)*exp(s1*t)+g(2)*exp(s2*t));
32:   if nargout > 2
33:     ah=real(s1*s1*g(1)*exp(s1*t)+ ...
34:        s2*s2*g(2)*exp(s2*t));
35:   end
36: else
37:   s=-c/(2*m);
```

```
38:    g1=y0; g2=v0-s*g1; yh=(g1+g2*t).*exp(s*t);
39:    if nargout > 2
40:        ah=real(s*(2*g2+s*g1+s*g2*t).*exp(s*t));
41:    end
42: end
```

6.2.3 GENERAL PROGRAM TO CONSTRUCT FOURIER EXPANSIONS

The final example in this chapter is a program to compute Fourier coefficients of general real valued functions and to display series with varying numbers of terms so that a user can see how rapidly such series converge. Since a truncated Fourier series is a continuous differentiable function, it cannot perfectly represent a discontinuous function such as a square wave. At points where jump discontinuities occur, Fourier series approximations oscillate [16]. The same behavior occurs less seriously at points of slope discontinuity. Surprisingly, adding more terms does not cure the problem at jump discontinuities. The behavior, known as Gibbs phenomenon, produces approximations which overshoot the function on either side of the discontinuity. Illustrations of this behavior appear below.

A program was written to expand real functions of arbitrary period using Fourier series approximations defined by applying the FFT to data values equally spaced in time over a period. The function can either be piecewise linear or can be represented by a MATLAB M-file. For instance, a function varying like a sine curve with the bottom half cut off would be

```
function y=chopsine(x,period)
y=sin(pi*x/period).*(x<period)
```

The program consists of the following functions.

fouseris	main driver
sine	example for exact function input
lintrp	function for piecewise linear interpolation
fousum	sum a real valued Fourier series
read	reads several data items on one line

Comments within the program illustrate how to input data interactively. Details of different input options can be found by executing the program.

Let us see how well the FFT approximates a function of period

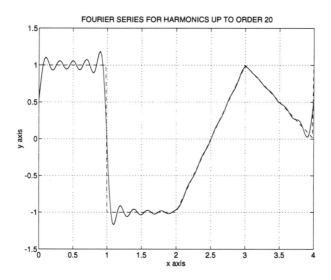

FIGURE 6.8. Fourier Series for Harmonics up to Order 20

3 defined by piecewise linear interpolation through (x, y) values of $(0,1)$, $(1,1)$, $(1,-1)$, $(2,-1)$, $(3,1)$, and $(4,0)$. The function has jump discontinuities at $x = 0$, $x = 1$, and $x = 4$. A slope discontinuity also occurs at $x = 3$. Program results using a twenty-term approximation appear in Figure 6.8. Results produced by 100- and 250-term series plotted near $x = 1$ are shown in Figures 6.9 and 6.10. Clearly, adding more terms does not eliminate the oscillation. However, the oscillation at a jump discontinuity can be reduced with the Lanczos smoothing procedure. Results for a series of 250 terms smoothed over an interval equal to the period times 0.01 appears in Figure 6.11. The oscillation is reduced at the cost of replacing the infinite slope at a discontinuity point by a steep slope of fifty-to-one for this case. Figure 6.12 shows a plot produced using an exact function definition as indicated in the second program execution. The reader may find it instructive to investigate how well Fourier series converge by running the program for other function choices.

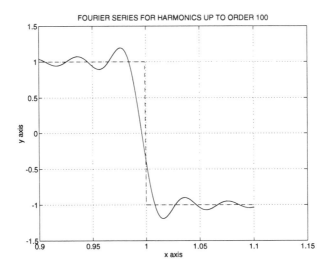

FIGURE 6.9. Fourier Series for Harmonics up to Order 100

FIGURE 6.10. Fourier Series for Harmonics up to Order 250

FIGURE 6.11. Smoothed Fourier Series for Harmonics to Order 250

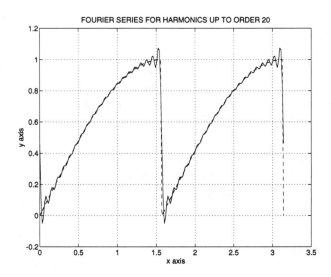

FIGURE 6.12. Exact Function Example for Harmonics up to Order 20

MATLAB EXAMPLE

6.2.3.1 Output for Piecewise Linear Example

```
>> fouseris

FOURIER SERIES EXPANSION FOR A PIECEWISE LINEAR OR
         ANALYTICALLY DEFINED FUNCTION

Input the period of the function
 ? > 4

Input the number of data points to define the function
by piecewise linear interpolation (input a zero if the
function is defined analytically by the user).
 ? > 6

Input the x,y values one pair per line
 ? > 0,1
 ? > 1,1
 ? > 1,-1
 ? > 2,-1
 ? > 3,1
 ? > 4,0

To plot the series input xmin, xmax, and the highest
harmonic not exceeding 255 (input 0,0,0 to stop)
(Use a negative harmonic number to save your graph)
 ? > 0,4,20

To plot the series smoothed over a fraction of the
period, input the smoothing fraction
(give 0.0 for no smoothing).
 ? > 0

Press RETURN to continue

To plot the series input xmin, xmax, and the highest
harmonic not exceeding 255 (input 0,0,0 to stop)
(Use a negative harmonic number to save your graph)
 ? > 0,0,0
```

6.2.3.2   Output for Analytically Defined Example

```
>> fouseris
```

FOURIER SERIES EXPANSION FOR A PIECEWISE LINEAR OR
        ANALYTICALLY DEFINED FUNCTION

Input the period of the function
 ? > pi/2

Input the number of data points to define the function
by piecewise linear interpolation (input a zero if the
function is defined analytically by the user).
 ? > 0

Select the method used for exact function definition:

1 <=> Use an existing function with syntax defined by
the following example:

```
function y=sine(x,period)
%
% y=sine(x,period)
% ~~~~~~~~~~~~~~~~~
% This function specifies all or part of
% a sine wave.
%
%    x         - vector of argument values
%    period  - period of the function
%    y         - vector of function values
%
% User m functions called:   none
%------------------------------------------------
y=sin(rem(x,period));
```

or

2 <=> Use a one-line character string definition
involving argument x and period p. For example a sine
wave with the bottom cut off would be defined by:

```
sin(x*2*pi/p).*(x<p/2)
```

1 or 2 ? > 1

Enter the name of your function
 ? > sine

To plot the series input xmin, xmax, and the highest
harmonic not exceeding 255 (input 0,0,0 to stop)
(Use a negative harmonic number to save your graph)
 ? > 0,pi,-20

To plot the series smoothed over a fraction of the
period, input the smoothing fraction
(give 0.0 for no smoothing).
 ? > 0

Give a file name to save the current graph >
exactplt

Press RETURN to continue

To plot the series input xmin, xmax, and the highest
harmonic not exceeding 255 (input 0,0,0 to stop)
(Use a negative harmonic number to save your graph)
 ? > 0,0,0

6.2.3.3   Script File fouseris

```
 1: % Example: fouseris
 2: % ~~~~~~~~~~~~~~~~~~
 3: % This program illustrates the convergence rate
 4: % of Fourier series approximations derived by
 5: % applying the FFT to a general function which
 6: % may be specified either by piecewise linear
 7: % interpolation in a data table or by
 8: % analytical definition in a function given by
 9: % the user. The linear interpolation model
10: % permits inclusion of jump discontinuities.
11: % Series having varying numbers of terms can
12: % be graphed to demonstrate Gibbs phenomenon
13: % and to show how well the truncated Fourier
14: % series represents the original function.
15: % Provision is made to plot the Fourier series
16: % of the original function or a smoothed
17: % function derived by averaging the original
18: % function over an arbitrary fraction of the
19: % total period.
20: %
21: % User m functions required:
22: %     fousum, lintrp, read, sine, genprint,
23: %     removfil
24: %--------------------------------------------------
25:
26: % The following parameters control the number
27: % of fft points used and the number of points
28: % used for graphing.
29: nft=512; ngph=200; nmax=int2str(nft/2-1);
30:
31: fprintf('\nFOURIER SERIES EXPANSION FOR')
32: fprintf(' A PIECEWISE LINEAR OR')
33: fprintf('\n           ANALYTICALLY DEFINED ')
34: fprintf('FUNCTION\n')
35:
36: fprintf('\nInput the period of the function')
37: period=input('? > ');
38: xfc=(period/nft)*(0:nft-1)';
39: fprintf('\nInput the number of data ')
40: fprintf('points to define the function')
41: fprintf('\nby piecewise linear ')
42: fprintf('interpolation (input a zero if the')
```

```
43: fprintf('\nfunction is defined analytically')
44: fprintf(' by the user).')
45: nd=input('? > ');
46: if nd > 0, xd=zeros(nd,1); yd=xd;
47:   fprintf('\nInput the x,y values one ')
48:   fprintf('pair per line')
49:   for j=1:nd
50:     [xd(j),yd(j)]=read('? > ');
51:   end
52:
53: % Use nft interpolated data points to
54: % compute the fft
55:   yfc=lintrp(xfc,xd,yd); c=fft(yfc);
56: else
57:   fprintf('\nSelect the method used for ')
58:   fprintf('exact function definition:\n')
59:   fprintf('\n1 <=> Use an existing function ')
60:   fprintf('with syntax defined by')
61:   fprintf('\nthe following example:\n')
62:   type sine.m
63:   fprintf('or\n')
64:   fprintf('\n2 <=> Use a one-line character')
65:   fprintf(' string definition involving')
66:   fprintf('\nargument x and period p. For')
67:   fprintf(' example a sine wave with the')
68:   fprintf('\nbottom cut off would be ')
69:   fprintf('defined by: ')
70:   fprintf('sin(x*2*pi/p).*(x<p/2)\n')
71:   nopt=input('1 or 2 ? > ');
72:   if nopt == 1
73:     fprintf('\nEnter the name of your ')
74:     fprintf('function')
75:     fnam=input('? > ','s');
76:     yfc=feval(fnam,xfc,period); c=fft(yfc);
77:   else
78:     fprintf('\nInput the one-line definition')
79:     fprintf(' in terms of x and p')
80:     strng=input('? > ','s');
81:     x=xfc; p=period;
82:     yfc=eval(strng); c=fft(yfc);
83:   end
84: end
85:
86: end
87: while 1
```

```
88:    fprintf('\nTo plot the series input xmin,')
89:    fprintf(' xmax, and the highest')
90:    fprintf(['\nharmonic not exceeding ', ...
91:              nmax,' (input 0,0,0 to stop)'])
92:    fprintf('\n(Use a negative harmonic number')
93:    fprintf(' to save your graph)')
94:    [xl,xu,nh]=read('? > ');
95:    if norm([xl,xu,nh])==0, break; end
96:    pltsav=(nh < 0); nh=abs(nh);
97:    xtmp=xl+((xu-xl)/ngph)*(0:ngph);
98:    fprintf('\nTo plot the series smoothed ')
99:    fprintf('over a fraction of the')
100:   fprintf('\nperiod, input the smoothing ')
101:   fprintf('fraction')
102:   fprintf('\n(give 0.0 for no smoothing).')
103:   alpha=input('? > ');
104:   yfou=fousum(c,xtmp,period,nh,alpha);
105:   xxtmp=xtmp; idneg=find(xtmp<0);
106:   xng=abs(xtmp(idneg));
107:   xxtmp(idneg)=xxtmp(idneg)+ ...
108:                 period*ceil(xng/period);
109:   if nd>0
110:     yexac=lintrp(rem(xxtmp,period),xd,yd);
111:   else
112:     if nopt == 1
113:       yexac=feval(fnam,xtmp,period);
114:     else
115:       x=xxtmp; yexac=eval(strng);
116:     end
117:   end
118:   in=int2str(nh);
119:   if alpha == 0
120:     titl=['FOURIER SERIES FOR HARMONICS ' ...
121:           'UP TO ORDER ',in];
122:   else
123:     titl=['SMOOTHED FOURIER SERIES FOR ' ...
124:           'HARMONICS UP TO ORDER ',in];
125:   end
126:   clf;
127:   plot(xtmp,yfou,'-w',xtmp,yexac,'--w');
128:   ylabel('y axis'); xlabel('x axis');
129:   title(titl); grid
130:   input('Press RETURN to continue ','s');
131:   if pltsav
132:     filnam=input(['Give a file name to ' ...
```

```
133:           'save the current graph > ? '],'s');
134:     if length(filnam) > 0
135:       genprint(filnam);
136:     end
137:   end
138: end
```

6.2.3.4 Function sine

```
1: function y=sine(x,period)
2: %
3: % y=sine(x,period)
4: % ~~~~~~~~~~~~~~~~~
5: % This function specifies all or part
6: % of a sine wave.
7: %
8: %   x        - vector of argument values
9: %   period  - period of the function
10: %   y        - vector of function values
11: %
12: % User m functions called:  none
13: %-----------------------------------------------
14:
15: y=sin(rem(x,period));
```

6.2.3.5 Function fousum

```
1: function yreal=fousum(c,x,period,k,alpha)
2: %
3: % yreal = fousum(c,x,period,k,alpha)
4: % ~~~~~~~~~~~~~~~~~~~~~~~~~~~~~~~~~~~~
5: % Sum the Fourier series of a real
6: % valued function.
7: %
8: %   x        - The vector of real values at
9: %              which the series is evaluated.
10: %   c        - A vector of length n containing
11: %              Fourier coefficients output by
12: %              the fft function
13: %   period - The period of the function
14: %   k        - The highest harmonic used in
```

```
15: %              the Fourier sum.  This must
16: %              not exceed n/2-1
17: %    alpha  - If this parameter is nonzero,
18: %              the Fourier coefficients are
19: %              replaced by those of a function
20: %              obtained by averaging the
21: %              original function over alpha
22: %              times the period
23: %    yreal  - The real valued Fourier sum
24: %              for argument x
25: %
26: % The Fourier coefficients c must have been
27: % computed using the fft function which
28: % transforms the vector [y(1),...,y(n)] into
29: % an array of complex Fourier coefficients
30: % which have been multiplied by n and are
31: % arranged in the order:
32: %
33: %    [c(0),c(1),...,c(n/2-1),c(n/2),
34: %              c(-n/2+1),...,c(-1)].
35: %
36: % The coefficient c(n/2) cannot be used
37: % since it is actually the sum of c(n/2) and
38: % c(-n/2). For a particular value of n, the
39: % highest usable harmonic is n/2-1.
40: %
41: % User m functions called:  none
42: %-------------------------------------------------
43:
44: x=x(:); n=length(c);
45: if nargin <4, k=n/2-1; alpha=0; end
46: if nargin <5, alpha=0; end
47: if nargin <3, period=2*pi; end
48: L=period/2; k=min(k,n/2-1); th=(pi/L)*x;
49: i=sqrt(-1); z=exp(i*th);
50: y=c(k+1)*ones(size(th)); pa=pi*alpha;
51: if alpha > 0
52:   jj=(1:k)';
53:   c(jj+1)=c(jj+1).*sin(jj*pa)./(jj*pa);
54: end
55: for j=k:-1:2
56:   y=c(j)+y.*z;
57: end
58: yreal=real(c(1)+2*y.*z)/n;
```

7

Dynamic Response of Linear Second Order Systems

7.1 Solving the Structural Dynamics Equations for Periodic Applied Forces

The dynamics of a linear structure subjected to periodic forces obeys the matrix differential equation

$$M\ddot{X} + C\dot{X} + KX = F(t)$$

with initial conditions

$$X(0) = D_0 \qquad \dot{x}(0) = V_0$$

The solution vector $X(t)$ has dimension n and M, C, and K are real square matrices of order n. The mass matrix, M, the damping matrix, C, and the stiffness matrix, K, are all real. The forcing function $F(t)$, assumed to be real and having period L, can be approximated by a finite trigonometric series as

$$F(t) = \sum_{k=-N}^{N} c_k e^{\imath \omega_k t} \qquad \imath = \sqrt{-1} \qquad \omega_k = 2\pi k/L.$$

The Fourier coefficients c_k are vectors which can be computed using the FFT. The fact that $F(t)$ is real also implies that $c_{-k} = \mathbf{conj}(c_k)$ and, therefore,

$$F(t) = c_0 + 2\,\mathbf{real}\left(\sum_{k=1}^{n} c_k e^{\imath \omega_k t}\right)$$

The solution of the differential equation is naturally resolvable into two distinct parts. The first is the so called particular or forced response which is periodic and has the same general mathematical form as the forcing function. Hence, we write

$$X_p = \sum_{k=-n}^{n} X_k e^{\imath \omega_k t} = X_0 + 2\,\mathbf{real}\left(\sum_{k=1}^{n} X_k e^{\imath \omega_k t}\right)$$

Substituting this series into the differential equation and matching coefficients of $e^{\imath \omega_k t}$ on both sides yields

$$X_k = (K - \omega_k^2 M + \imath \omega_k C)^{-1} c_k$$

The particular solution satisfies initial conditions given by

$$X_p(0) = X_0 + 2 \textbf{ real} \left(\sum_{k=1}^{n} c_k \right)$$

and

$$\dot{X}_p(0) = 2 \textbf{ real} \left(\sum_{k=1}^{n} \imath \omega_k c_k \right)$$

Since these conditions usually will not equal the desired values, the particular solution must be combined with what is called the homogeneous or transient solution X_k

$$M \ddot{X}_h + c \dot{X}_h + K X_h = 0$$

$$X_h(0) = D_0 - X_p(0) \qquad \dot{X}_h(0) = V_0 - \dot{X}_p(0)$$

The homogeneous solution can be constructed by reducing the original differential equation to first order form. Let Z be the vector of dimension $2n$ which is the concatenation of X and $\dot{X} = V$. Hence, $Z = [X; V]$ and the original equation of motion is

$$\frac{dZ}{dt} = AZ + P(t)$$

where

$$A = \begin{bmatrix} 0 & I \\ -M^{-1}K & -M^{-1}C \end{bmatrix} \qquad P = \begin{bmatrix} 0 \\ m^{-1}F \end{bmatrix}$$

The homogeneous differential equation resulting when $P = 0$ can be solved in terms of the eigenvalues and eigenvectors of matrix A. If we know eigenvalues λ_j and eigenvectors U_j satisfying

$$A U_j = \lambda_j U_j \qquad 1 \le j \le 2n$$

then the homogeneous solution can be written as

$$Z = \sum_{j=1}^{2n} z_j U_j e^{\imath \omega_j t}$$

The weighting coefficients z_j are computed to satisfy the desired initial conditions which require

$$\left[\begin{array}{cccc} U_1, & U_2, & \cdots, & U_{2n} \end{array} \right] \left[\begin{array}{c} z_1 \\ \vdots \\ z_{2n} \end{array} \right] = \left[\begin{array}{c} X_0 - X_p(0) \\ V_0 - \dot{X}_p(0) \end{array} \right]$$

We solve this system of equations for $z_1, \cdots, z_{2n}$ and replace each U_j by $z_j U_j$. Then the homogeneous solution is

$$X_h = \sum_{j=1}^{n} U_j(1:n) e^{\lambda_j t}$$

where $U_j(1:n)$ means we take only the first n elements of column j.

In most practical situations, matrix C is nonzero and the eigenvalues $\lambda_1, \cdots, \lambda_{2n}$ have negative real parts. Then the exponential terms $e^{\lambda_j t}$ all decay rapidly, which explains the name, the transient solution, some authors give to X_h. In other cases, where the damping matrix C is zero, the eigenvalues λ_j are typically purely imaginary, and the homogeneous solution does not die out. In either instance, it is often customary in practical situations to ignore the homogeneous solution because it is usually small when compared to the contribution of the particular solution.

7.1.1 APPLICATION TO OSCILLATIONS OF A VERTICALLY SUSPENDED CABLE

Let us solve the problem of small transverse vibrations of a vertically suspended cable. This system illustrates nicely how the natural frequencies and mode shapes of a linear system can be combined to satisfy general initial conditions on position and velocity.

The cable in Figure 7.1 is idealized as a series of n rigid links connected at frictionless joints. Two vectors, consisting of link lengths $[\ell_1, \ell_2, \cdots, \ell_n]$ and masses $[m_1, m_2, \cdots, m_n]$ lumped at the joints, characterize the system properties. The accelerations in the vertical direction will be negligibly small compared to transverse accelerations, because the transverse displacements are small. Consequently, the tension in the chain will remain close to the static equilibrium value. This means the tension in link i is

$$T_i = gb_i \qquad b_i = \sum_{j=i}^{n} m_j$$

We assume that the transverse displacement y_i for mass m_i is small compared to the total length of the cable. A free body diagram for mass i is shown in Figure 7.2. The small deflection angles are related to the transverse deflections by $\theta_{i+1} = (y_{i+1} - y_i)\, \ell_{i+1}$ and $\theta_i = (y_i - y_{i-1})/\ell_i$. Summation of forces shows that the horizontal acceleration is governed by

$$
\begin{aligned}
m_i \ddot{y}_i &= g(b_{i+1}/\ell_{i+1})(y_{i+1} - y_i) - g(b_i/\ell_i)(y_i - y_{i-1}) \\[2mm]
&= g(b_i/\ell_i)y_{i-1} - g(b_i/\ell_i + b_{i+1}/\ell_{i+1})y_i + g(b_{i+1}/\ell_{i+1})y_{i+1}
\end{aligned}
$$

In matrix form this equation is

$$M\ddot{Y} + KY = 0$$

where M is a diagonal matrix of mass coefficients and K is a symmetric tridiagonal matrix. The natural modes of free vibration are dynamical states where each element of the system simultaneously moves with harmonic motion of the same frequency. This means we seek motions of the form $Y = U\cos(\omega t)$, or equivalently $Y = U\sin(\omega t)$, which implies

$$KU_j = \lambda_j MU_j \qquad \lambda_j = \omega_j^2 \qquad 1 \le j \le n$$

Solving the eigenvalue problem $(M^{-1}K)U = \lambda U$ gives the natural frequencies $\omega_1, \cdots, \omega_n$ and the modal vectors $U_1, \cdots, U_n$. The response to general initial conditions is then obtained by superposition of the component modes. We write

$$Y = \sum_{j=1}^{n} \cos(\omega_j t)U_j c_j + \sin(\omega_j t)U_j d_j/\omega_j$$

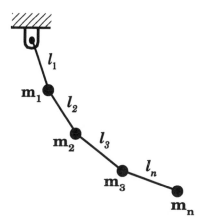

FIGURE 7.1. Transverse Cable Vibration

where coefficients $c_1, \cdots, c_n$ and $d_1, \cdots, d_n$ (not to be confused with Fourier coefficients) are determined from the initial conditions as

$$
\left[\begin{array}{ccc} U_1, & \cdots, & U_n \end{array}\right] \left[\begin{array}{c} c_1 \\ \vdots \\ c_n \end{array}\right] = Y(0) \qquad c = U^{-1}Y(0)
$$

$$
\left[\begin{array}{ccc} U_1, & \cdots, & U_n \end{array}\right] \left[\begin{array}{c} d_1 \\ \vdots \\ d_n \end{array}\right] = \dot{Y}(0) \qquad d = U^{-1}\dot{Y}(0)
$$

and the system response is complete.

The following program determines the cable response for general initial conditions. The natural frequencies and mode shapes are computed along with an animation of the motion.

The cable motion produced when an initially vertical system is given the same initial transverse velocity for all masses was studied. Graphical results of the analysis appear in Figures 7.3 through 7.6. The surface plot in Figure 7.3 shows the cable deflection pattern in terms of longitudinal position and time. Figure 7.4 shows the deflection pattern at two times. Figure 7.5 traces the motion of the middle and the free end. At $t = 1$, the wave propagating downward from the support point is about halfway down the cable. By $t = 2$, the wave has reached the free end and the cable

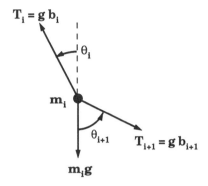

FIGURE 7.2. Forces on i'th Mass

is about to swing back. Finally, traces of cable positions during successive stages of motion appear in Figure 7.6.

SURFACE SHOWING CABLE DEFLECTION

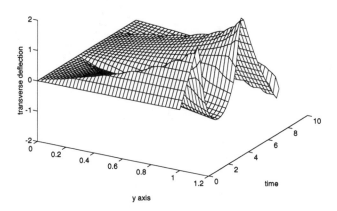

FIGURE 7.3. Surface Showing Cable Deflection

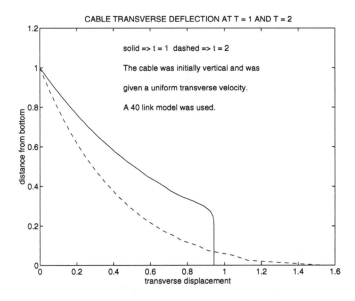

FIGURE 7.4. Cable Transverse Deflection at $T = 1$ and $T = 2$

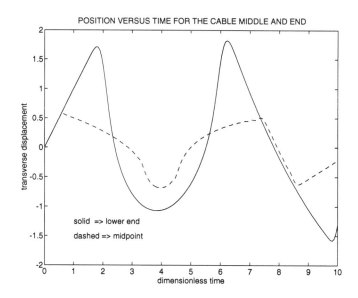

FIGURE 7.5. Position Versus Time for the Cable Middle and End

FIGURE 7.6. Trace of Cable Motion

MATLAB EXAMPLE

7.1.1.1   Script File cablinea

```
 1: % Example: cablinea
 2: % ~~~~~~~~~~~~~~~~~~
 3: % This program uses modal superposition to
 4: % compute the dynamic response of a cable
 5: % suspended at one end and free at the other.
 6: % The cable is given a uniform initial
 7: % velocity. Time history plots and animation
 8: % of the motion are provided.
 9: %
10: % User m functions required:
11: %      cablemk, udfrevib, removfil, canimate,
12: %      genprint
13: %-------------------------------------------------
14:
15: % Initialize graphics
16: hold off; axis('normal'); clf;
17:
18: % Set physical parameters
19: n=40; gravty=1.; masses=ones(n,1)/n;
20: lengths=ones(n,1)/n;
21:
22: % Obtain mass and stiffness matrices
23: [m,k]=cablemk(masses,lengths,gravty);
24:
25: % Assign initial conditions & time limit
26: % for solution
27: dsp=zeros(n,1); vel=ones(n,1);
28: tmin=0; tmax=10; ntim=40;
29:
30: % Compute the solution by modal superposition
31: [t,u,modvc,natfrq]=...
32:     udfrevib(m,k,dsp,vel,tmin,tmax,ntim);
33:
34: % Interpret results graphically
35: nt1=sum(t<=tmin); nt2=sum(t<=tmax);
36: u=[zeros(ntim,1),u];
37: y=cumsum(lengths); y=[0;y(:)];
38:
39: % Plot deflection surface
40: clf; mesh(y,t,u);
```

224

```
41: xlabel('y axis'); ylabel('time');
42: zlabel('transverse deflection');
43: title('SURFACE SHOWING CABLE DEFLECTION');
44: view([30,30])
45: disp('Press RETURN to continue'); pause
46: genprint('surface');
47:
48: % Show deflection configuration at two times
49: % Use closer time increment than was used
50: % for the surface plots.
51: mtim=4*ntim;
52: [tt,uu,modvc,natfrq]=...
53:    udfrevib(m,k,dsp,vel,tmin,tmax,mtim);
54: uu=[zeros(mtim,1),uu];
55: tp1=.1*tmax; tp2=.2*tmax;
56: s1=num2str(tp1); s2=num2str(tp2);
57: np1=sum(tt<=tp1); np2=sum(tt<=tp2);
58: u1=uu(np1,:);u2=uu(np2,:);
59: yp=flipud(y(:)); ym=max(yp);
60: plot(u1,yp,'-w',u2,yp,'--w')
61: ylabel('distance from bottom')
62: xlabel('transverse displacement');
63: xm=.1*max([u1(:);u2(:)]);
64: title(['CABLE TRANSVERSE DEFLECTION ' ...
65:        'AT T = ',s1,' AND T = ',s2])
66: text(xm,.9*ym, ...
67:     ['solid => t = ',s1,'  dashed => t = ',s2])
68: text(xm,.8*ym, ...
69:     'The cable was initially vertical and was')
70: text(xm,.7*ym, ...
71:     'given a uniform transverse velocity.')
72: ntxt=int2str(n); n2=1+fix(n/2);
73: text(...
74: xm,.6*ym,['A ',ntxt,' link model was used.'])
75: disp('Press RETURN to continue'); pause
76: genprint('twoposn');
77:
78: % Plot time history for the middle and the end
79: plot(tt,uu(:,n2),'--w',tt,uu(:,n+1),'-w')
80: xlabel('dimensionless time')
81: ylabel('transverse displacement')
82: title(['POSITION VERSUS TIME FOR THE ' ...
83:        'CABLE MIDDLE AND END'])
84: text(.1*tmax,-.50,'solid  => lower end')
85: text(.1*tmax,-.75,'dashed => midpoint')
```

```
86: disp('Press RETURN to continue'); pause
87: genprint('2timhist');
88:
89: % Plot animation of motion history
90: canimate(y,u,t,0,.5*max(t))
91: genprint('motntrac');
```

7.1.1.2 Function cablemk

```
 1: function [m,k]=cablemk(masses,lngths,gravty)
 2: %
 3: % [m,k]=cablemk(masses,lngths,gravty)
 4: % ~~~~~~~~~~~~~~~~~~~~~~~~~~~~~~~~~~~~
 5: % Form the mass and stiffness matrices for
 6: % the cable.
 7: %
 8: % masses      - vector of masses
 9: % lngths      - vector of link lengths
10: % gravty      - gravity constant
11: % m,k         - mass and stiffness matrices
12: %
13: % User m functions called:  none.
14: %------------------------------------------------
15:
16: m=diag(masses);
17: b=flipud(cumsum(flipud(masses(:))))* ...
18:    gravty./lngths;
19: n=length(masses); k=zeros(n,n); k(n,n)=b(n);
20: for i=1:n-1
21:    k(i,i)=b(i)+b(i+1); k(i,i+1)=-b(i+1);
22:    k(i+1,i)=k(i,i+1);
23: end
```

7.1.1.3 Function udfrevib

```
 1: function [t,u,modvcs,natfrq]=...
 2:                 udfrevib(m,k,u0,v0,tmin,tmax,nt)
 3: %
 4: % [t,u,mdvc,natfrq]= ...
 5: %                 udfrevib(m,k,u0,v0,tmin,tmax,nt)
 6: % ~~~~~~~~~~~~~~~~~~~~~~~~~~~~~~~~~~~~~~~~~~~~~~~~
```

```
7: % This function computes undamped natural
8: % frequencies, modal vectors, and time response
9: % by modal superposition. The matrix
10: % differential equation and initial conditions
11: % are
12: %
13: %     m u'' + k u = 0,  u(0) = u0, u'(0) = v0
14: %
15: % m,k        - mass and stiffness matrices
16: % u0,v0      - initial position and velocity
17: %              vectors
18: % tmin,tmax  - time limits for solution
19: %              evaluation
20: % nt         - number of times for solution
21: % t          - vector of solution times
22: % u          - matrix with row j giving the
23: %              system response at time t(j)
24: % mdvc       - matrix with columns which are
25: %              modal vectors
26: % natfrq     - vector of natural frequencies
27: %
28: % User m functions called:  none.
29: %-----------------------------------------------
30:
31: % Call function eig to compute modal vectors
32: % and frequencies
33: [mdvc,w]=eig(m\k);
34: [w,id]=sort(diag(w)); w=sqrt(w);
35:
36: % Arrange frequencies in ascending order
37: mdvc=mdvc(:,id); z=mdvc\[u0(:),v0(:)];
38:
39: % Generate vector of equidistant times
40: t=tmin+(tmax-tmin)/(nt-1)*(0:nt-1);
41:
42: % Evaluate the displacement as a
43: % function of time
44: u=(mdvc*diag(z(:,1)))*cos(w*t)+...
45:   (mdvc*diag(z(:,2)./w))*sin(w*t);
46: t=t(:); u=u'; natfrq=w;
```

7.1.1.4   Function canimate

```
 1: function canimate(y,u,t,tmin,tmax,norub)
 2: %
 3: % canimate(y,u,t,tmin,tmax,norub)
 4: % ~~~~~~~~~~~~~~~~~~~~~~~~~~~~~~~~~
 5: % This function draws an animated plot of
 6: % data values stored in array u. The
 7: % different columns of u correspond to position
 8: % values in vector y. The successive rows of u
 9: % correspond to different times. Parameter
10: % tpause controls the speed of the animation.
11: %
12: % u          - matrix of values for which
13: %                animated plots of u versus y
14: %                are required
15: % y          - spatial positions for different
16: %                columns of u
17: % t          - time vector at which positions
18: %                are known
19: % tmin,tmax  - time limits for graphing of the
20: %                solution
21: % norub      - parameter which makes all
22: %                position images remain on the
23: %                screen. Only one image at a
24: %                time shows if norub is left out.
25: %                A new cable position appears each
26: %                time the user presses any key
27: %
28: % User m functions called:  none.
29: %-------------------------------------------------
30:
31: % Determine window limits
32: umin=min(u(:)); umax=max(u(:)); udif=umax-umin;
33: uavg=.5*(umin+umax);
34: ymin=min(y); ymax=max(y); ydif=ymax-ymin;
35: yavg=.5*(ymin+ymax);
36: ywmin=yavg-.55*ydif; ywmax=yavg+.55*ydif;
37: uwmin=uavg-.55*udif; uwmax=uavg+.55*udif;
38: n1=sum(t<=tmin); n2=sum(t<=tmax);
39: t=t(n1:n2); u=u(n1:n2,:);
40: u=fliplr (u); [ntime,nxpts]=size(u);
41:
42: hold off; cla; ey=0; eu=0; axis('square')
```

```
43: axis([uwmin,uwmax,ywmin,ywmax]);
44: axis off; hold on
45: title('TRACE OF CABLE MOTION')
46:
47: % If norub is input,
48: %   all images are left on the screen
49: if nargin < 6
50:   rubout = 1;
51: else
52:   rubout = 0;
53: end
54:
55: % Plot successive positions
56: for j=1:ntime
57:   ut=u(j,:); plot(ut,y,'w');
58:   fprintf('\nPress RETURN to continue, ')
59:   fprintf('Ctrl-C to exit\n');
60:   pause
61:
62:   % Erase image before next one appears
63:   if rubout, cla, end
64: end
65: pause
```

7.2   Direct Integration Methods

Using stepwise integration methods to solve the structural dynamics equation provides an alternative to frequency analysis methods. If we invert the mass matrix and save the result for later use, the n degree-of-freedom system can be expressed concisely as a first order system in $2n$ unknowns for a vector by $z = [x; v]$, where v is the time derivative of x. The system can be solved by applying the variable step-size differential equation integrator **ode45** to the following function:

```
function zdot = sdeq(t,z)
% global invmas_ damp_ stif_ forcname_
n=length(z)/2; x=z(1:n); v=z(n+1:2*n); fnam=forcname_;
zdot=[v;invmas_*(feval(fnam,t)-stif_*x-damp_*v)];
```

In this function, the inverted mass matrix has been stored in a global variable **invmas_**, the damping and stiffness matrices are in **damp_** and **stiff_**, and the forcing function name is a character string called **forcname_**. Although this approach is easy to implement, the resulting analysis can be very time consuming for large systems. Variable step integrators make adjustments to control stability and accuracy which often lead to very small integration steps. Consequently, alternative formulations employing fixed step-size are usually chosen. We will investigate two such algorithms derived from trapezoidal integration rules [6, 97]. The two fundamental integration formulas [22] needed are:

$$\int_a^b f(t)dt = \frac{h}{2}[f(a) + f(b)] - \frac{h^3}{12}f''(\epsilon_1) \qquad \left\{ \begin{array}{l} a \le \epsilon_1 \le b \\ h = (b - a) \end{array} \right.$$

and

$$\int_a^b f(t)dt = \frac{h}{2}[f(a) + f(b)] + \frac{h^2}{12}[f'(a) - f'(b)] +$$

$$\frac{h^5}{720}f^{(4)}(\epsilon_2) \qquad\qquad a \le \epsilon_2 \le b$$

The first formula, called the trapezoidal rule, gives a zero truncation error term when applied to a linear function. Similiarly, the second formula, called the trapezoidal rule with end correction, has a zero final term for a cubic integrand.

The idea is to multiply the differential equation by dt, integrate from t to $(t+h)$, and employ numerical integration formulas while observing that M, C, and K are constant matrices, or

$$M \int_t^{t+h} \dot{V} \, dt + C \int_t^{t+h} \dot{X} \, dt + K \int_t^{t+h} X \, dt = \int_t^{t+h} P(t) \, dt$$

and

$$\int_t^{t+h} \dot{X} \, dt = \int_t^{t+h} V \, dt$$

For brevity we utilize a notation characterized by $X(t) = X_0$, $X(t+h) = X_1$, $\tilde{X} = X_1 - X_0$. The trapezoidal rule immediately leads to

$$\left[M + \frac{h}{2}C + \frac{h^2}{4}K \right] \tilde{V} = \int P(t) dt -$$

$$h \left[CV_0 + K(X_0 + \frac{h}{2}V_0) \right] + O(h^3)$$

The last equation is a balance of impulse and momentum change involving the effective mass matrix

$$M_e = [M + \frac{h}{2}C + \frac{h^2}{4}K]$$

which can be inverted once and used repeatedly if the step-size is not changed.

To integrate the forcing function we can use the midpoint rule [22] which states

$$\int P(t) \, dt = hP(\frac{a+b}{2}) + O(h^3)$$

Solving for $\tilde{V}$ yields

$$\tilde{V} = \left[M + \frac{h}{2}C + \frac{h^2}{4}K \right]^{-1} \left[P(t + \frac{h}{2}) - CV_0 - \right.$$

$$\left. K(X_0 + \frac{h}{2}V_0) \right] h + O(h^3)$$

The velocity and position at $(t + h)$ are then computed as

$$V_1 = V_0 + \tilde{V} \qquad X_1 = X_0 + \frac{h}{2}[V_0 + V_1] + O(h^3)$$

A more accurate formula with truncation error of order h^5 can be developed from the extended trapezoidal rule. This leads to

$$M\tilde{V} + C\tilde{X} + K\left[\frac{h}{2}(\tilde{X} + 2X_0) - \frac{h^2}{12}\tilde{V}\right] = \int P(t)dt + O(h^5)$$

and

$$\tilde{X} = \frac{h}{2}[\tilde{V} + 2V_0] + \frac{h^2}{12}[\dot{V}_0 - \dot{V}_1] + O(h^5)$$

Multiplying the last equation by M and employing the differential equation to reduce the $\dot{V}_0 - \dot{V}_1$ terms gives

$$M\tilde{X} = \frac{h}{2}M[\tilde{V} + 2V_0] + \frac{h^2}{12}[-\tilde{P} + C\tilde{V} + K\tilde{X}] + O(h^5)$$

These results can be arranged into a single matrix equation to be solved for $\tilde{X}$ and $\tilde{V}$, or

$$\begin{bmatrix} -(\frac{h}{2}M + \frac{h^2}{12}C) & (M - \frac{h^2}{12}K) \\ \\ (M - \frac{h^2}{12}K) & (C + \frac{h}{2}K) \end{bmatrix} \begin{bmatrix} \tilde{V} \\ \\ \tilde{X} \end{bmatrix} =$$

$$\begin{bmatrix} hMV_0 + \frac{h^2}{12}(P_0 - P_1) \\ \\ \int Pdt - hKX_0 \end{bmatrix} + O(h^5)$$

A Gauss two-point formula [22] evaluates the force integral consistent with the desired error order so that

$$\int_t^{t+h} P(t)dt = \frac{h}{2}[P(t + \alpha h) + P(t + \beta h)] + O(h^5)$$

where $\alpha = \frac{3-\sqrt{3}}{6}$ and $\beta = \frac{3+\sqrt{3}}{6}$.

7.2.1 EXAMPLE ON CABLE RESPONSE BY DIRECT INTEGRATION

Functions implementing the last two algorithms appear in the following program which solves the previously considered cable dynamics example by direct integration. Questions of computational efficiency and numerical accuracy are examined for two different step-sizes. Figures 7.7 and 7.8 present solution times as multiples of the times needed for a modal response solution. The accuracy measures employed are described next. Note that the displacement response matrix has rows describing system positions at successive times. Consequently, a measure of the difference between approximate and exact solutions is given by the vector

```
error_vector = sqrt(sum(((x_aprox-x_exact).^2)'));
```

Typically this vector has small initial components (near $t = 0$) and larger components (near the final time). The error measure is compared for different integrators and time steps in the figures. Note that the fourth order integrator is more efficient than the second order integrator because a larger integration step can be taken without excessive accuracy loss. Using $h = 0.4$ for **mckde4i** achieved nearly the same accuracy as that given by **mckde2i** with $h = 0.067$. However, the computation time for **mckde2i** was several times as large as that for **mckde4i**.

In the past it has been traditional to use only second order methods for solving the structural dynamics equation. This may have been dictated by considerations on computer memory. Since workstations widely available today have relatively large memories and can invert a matrix of order two-hundred in a couple of seconds, it appears that use of high order integrators may gain in popularity.

The following computer program concludes our chapter on solution of linear, constant-coefficient matrix differential equations. The next chapter deals with integration of nonlinear problems.

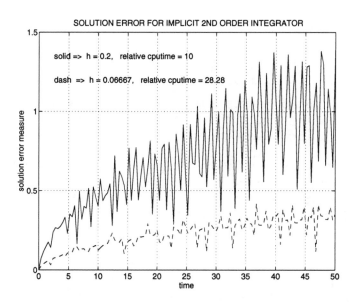

FIGURE 7.7. Solution Error for Implicit 2nd Order Integrator

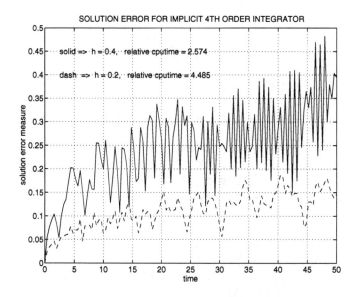

FIGURE 7.8. Solution Error for Implicit 4th Order Integrator

MATLAB Example

7.2.1.1   Script File deislner

```
 1: % Example:  deislner
 2: % ~~~~~~~~~~~~~~~~~~
 3: % Solution error for simulation of cable
 4: % motion using a second or a fourth order
 5: % implicit integrator.
 6: %
 7: % This program uses implicit second or fourth
 8: % order integrators to compute the dynamical
 9: % response of a cable which is suspended at
10: % one end and is free at the other end. The
11: % cable is given a uniform initial velocity.
12: % A plot of the solution error is given for
13: % two cases where approximate solutions are
14: % generated using numerical integration rather
15: % than modal response which is exact.
16: %
17: % User m functions required:
18: %    mckde2i, mckde4i, cablemk, udfrevib,
19: %    removfil, genprint
20: %-------------------------------------------------
21:
22: % Choose a model having twenty
23: % links of equal length
24: clear
25: axis('normal'); clf;
26: n=20; gravty=1.; n2=1+fix(n/2);
27: masses=ones(n,1)/n; lengths=ones(n,1)/n;
28:
29: % Set flag to control the choice
30: % of integrator. Use 1 for the second
31: % order integrator or 2 for the fourth
32: % order integrator.
33: solntyp=1;
34:
35: % First generate the exact solution by
36: % modal superposition
37: [m,k]=cablemk(masses,lengths,gravty);
38: c=zeros(size(m));
39: dsp=zeros(n,1); vel=ones(n,1);
40: t0=0; tfin=50; ntim=126;
```

```
41: tcpmr=cputime;
42: [tmr,xmr]=udfrevib(m,k,dsp,vel,t0,tfin,ntim);
43: tcpmr=cputime-tcpmr;
44:
45: % Then compute approximate results obtained
46: % by direct integration
47: if solntyp ==1
48:    incout=06; h=(tfin-t0)/(incout*(ntim-1));
49:    Incout=02; H=(tfin-t0)/(Incout*(ntim-1));
50: else
51:    incout=02; h=(tfin-t0)/(incout*(ntim-1));
52:    Incout=01; H=(tfin-t0)/(Incout*(ntim-1));
53: end
54:
55: % Results for small timestep
56: if solntyp == 1
57:    typ='2ND';
58:    [t2i,x2i,tcp2i] = ...
59:    mckde2i(m,c,k,t0,dsp,vel,tfin,h,incout);
60: else
61:    typ='4TH';
62:    [t2i,x2i,tcp2i] = ...
63:    mckde4i(m,c,k,t0,dsp,vel,tfin,h,incout);
64: end
65:
66: % Results for a timestep twice as large
67: % as the first choice
68: if solntyp ==1
69:    [T2i,X2i,Tcp2i] = ...
70:    mckde2i(m,c,k,t0,dsp,vel,tfin,H,Incout);
71: else
72:    [T2i,X2i,Tcp2i] = ...
73:    mckde4i(m,c,k,t0,dsp,vel,tfin,H,Incout);
74: end
75:
76: % Compute error measures of solution accuracy
77: % and plot values.
78: d=x2i-xmr; er=sqrt(sum((d.*d)'));
79: D=X2i-xmr; Er=sqrt(sum((D.*D)'));
80: plot(t2i,er,'w--',t2i,Er,'w-'); xlabel('time')
81: ylabel('solution error measure')
82: title(['SOLUTION ERROR FOR IMPLICIT ',...
83:        typ,' ORDER INTEGRATOR'])
84: hh=gca;
85: v(1:2)=get(hh,'XLim'); v(3:4)=get(hh,'YLim');
```

```
86: r=tcp2i/tcpmr; R=Tcp2i/tcpmr;
87: str=['solid => h = ',num2str(H),...
88:      ',    relative cputime = ',  num2str(R)];
89: Str=['dash  => h = ',num2str(h),...
90:      ',    relative cputime = ',  num2str(r)];
91: text(v(1)+(v(2)-v(1))*.05, ...
92:      v(3)+(v(4)-v(3))*.9,str);
93: text(v(1)+(v(2)-v(1))*.05, ...
94:      v(3)+(v(4)-v(3))*.8,Str);
95: grid
96:
97: % Save graphical results
98: if solntyp ==1
99:   genprint('deislne2');
100: else
101:   genprint('deislne4');
102: end
```

7.2.1.2   Function mckde2i

```
1: function [t,x,tcp] = ...
2:     mckde2i(m,c,k,t0,x0,v0,tmax,h,incout,forc)
3: %
4: % [t,x,tcp]= ...
5: %     mckde2i(m,c,k,t0,x0,v0,tmax,h,incout,forc)
6: % ~~~~~~~~~~~~~~~~~~~~~~~~~~~~~~~~~~~~~~~~~~~~~~
7: % This function uses a second order implicit
8: % integrator % to solve the matrix differential
9: % equation
10: %           m x'' + c x' + k x = forc(t)
11: % where m,c, and k are constant matrices and
12: % forc is an externally defined function.
13: %
14: % Input:
15: % ------
16: % m,c,k    mass, damping and stiffness matrices
17: % t0       starting time
18: % x0,v0    initial displacement and velocity
19: % tmax     maximum time for solution evaluation
20: % h        integration stepsize
21: % incout   number of integration steps between
22: %          successive values of output
23: % forc     externally defined time dependent
```

```
24: %            forcing function. This parameter
25: %            should be omitted if no forcing
26: %            function is used.
27: %
28: % Output:
29: %  -------
30: % t            time vector going from t0 to tmax
31: %             in steps of
32: % x            h*incout to yield a matrix of
33: %             solution values such that row j
34: %             is the solution vector at time t(j)
35: % tcp          computer time for the computation
36: %
37: % User m functions called:  none.
38: %------------------------------------------------
39:
40: if (nargin > 9); force=1; else, force=0; end
41: tcp=cputime; hbig=h*incout;
42: t=(t0:hbig:tmax)'; n=length(t);
43: ns=(n-1)*incout; ts=t0+h*(0:ns)';
44: xnow=x0(:); vnow=v0(:);
45: nvar=length(x0);
46: jrow=1; jstep=0; h2=h/2;
47:
48: % Form the inverse of the effective
49: % stiffness matrix
50: mnv=h*inv(m+h2*(c+h2*k));
51:
52: % Initialize the output matrix for x
53: x=zeros(n,nvar); x(1,:)=xnow';
54: zroforc=zeros(length(x0),1);
55:
56: % Main integration loop
57: for j=1:ns
58:    tj=ts(j);tjh=tj+h2;
59:    if force
60:       dv=feval(forc,tjh);
61:    else
62:       dv=zroforc;
63:    end
64:    dv=mnv*(dv-c*vnow-k*(xnow+h2*vnow));
65:    vnext=vnow+dv;xnext=xnow+h2*(vnow+vnext);
66:    jstep=jstep+1;
67:    if jstep == incout
68:       jstep=0; jrow=jrow+1; x(jrow,:)=xnext';
```

```
69:    end
70:    xnow=xnext; vnow=vnext;
71: end
72: tcp=cputime-tcp;
```

7.2.1.3 Function mckde4i

```
 1: function [t,x,tcp] = ...
 2:        mckde4i(m,c,k,t0,x0,v0,tmax,h,incout,forc)
 3: %
 4: % [t,x,tcp]= ...
 5: %    mckde4i(m,c,k,t0,x0,v0,tmax,h,incout,forc)
 6: % ~~~~~~~~~~~~~~~~~~~~~~~~~~~~~~~~~~~~~~~~~~~~~~~~~
 7: % This function uses a fourth order implicit
 8: % integrator with fixed stepsize to solve the
 9: % matrix differential equation
10: %            m x'' + c x' + k x = forc(t)
11: % where m,c, and k are constant matrices and
12: % forc is an externally defined function.
13: %
14: % Input:
15: % ------
16: % m,c,k    mass, damping and stiffness matrices
17: % t0       starting time
18: % x0,v0    initial displacement and velocity
19: % tmax     maximum time for solution evaluation
20: % h        integration stepsize
21: % incout   number of integration steps between
22: %          successive values of output
23: % forc     externally defined time dependent
24: %          forcing function. This parameter
25: %          should be omitted if no forcing
26: %          function is used.
27: %
28: % Output:
29: % -------
30: % t        time vector going from t0 to tmax
31: %          in steps of h*incout
32: % x        matrix of solution values such
33: %          that row j is the solution vector
34: %          at time t(j)
35: % tcp      computer time for the computation
36: %
```

```
37: % User m functions called:  none.
38: %--------------------------------------------
39:
40: if nargin > 9, force=1; else, force=0; end
41: tcp=cputime; hbig=h*incout; t=(t0:hbig:tmax)';
42: n=length(t); ns=(n-1)*incout; nvar=length(x0);
43: jrow=1; jstep=0; h2=h/2; h12=h*h/12;
44:
45: % Form the inverse of the effective stiffness
46: % matrix for later use.
47:
48: m12=m-h12*k;
49: mnv=inv([[(-h2*m-h12*c),m12];
50:        [m12,(c+h2*k)]]);
51:
52: % The forcing function is integrated using a
53: % 2 point Gauss rule
54: r3=sqrt(3); b1=h*(3-r3)/6; b2=h*(3+r3)/6;
55:
56: % Initialize output matrix for x and other
57: % variables
58: xnow=x0(:); vnow=v0(:);
59: tnow=t0; zroforc=zeros(length(x0),1);
60:
61: if force
62:    fnow=feval(forc,tnow);
63: else
64:    fnow=zroforc;
65: end
66: x=zeros(n,nvar); x(1,:)=xnow'; fnext=fnow;
67:
68: % Main integration loop
69: for j=1:ns
70:    tnow=t0+(j-1)*h; tnext=tnow+h;
71:    if force
72:      fnext=feval(forc,tnext);
73:      di1=h12*(fnow-fnext);
74:      di2=h2*(feval(forc,tnow+b1)+ ...
75:            feval(forc,tnow+b2));
76:      z=mnv*[(di1+m*(h*vnow)); (di2-k*(h*xnow))];
77:      fnow=fnext;
78:    else
79:      z=mnv*[m*(h*vnow); -k*(h*xnow)];
80:    end
81:    vnext=vnow + z(1:nvar);
```

```
82:    xnext=xnow + z((nvar+1):2*nvar);
83:    jstep=jstep+1;
84:
85:    % Save results every incout steps
86:    if jstep == incout
87:       jstep=0; jrow=jrow+1; x(jrow,:)=xnext';
88:    end
89:
90:    % Update quantities for next step
91:    xnow=xnext; vnow=vnext; fnow=fnext;
92: end
93: tcp=cputime-tcp;
```

8

Integration of Nonlinear Initial Value Problems

8.1 General Concepts on Numerical Integration of Nonlinear Matrix Differential Equations

Methods for solving differential equations numerically are one of the most valuable analysis tools now available. Less expensive computer power and more friendly software are stimulating wider use of digital simulation methods. At the same time, intelligent use of numerically integrated solutions requires appreciation of inherent limitations of the techniques employed. The present chapter discusses the widely used Runge-Kutta method and applies it to some specific examples.

When physical systems are described by mathematical models, it is common that various system parameters are only known approximately. For example, to predict the response of a building undergoing earthquake excitation, simplified formulations may be necessary to handle the elastic and frictional characteristics of the soil and the building. Our observation that simple models are used often to investigate behavior of complex systems does not necessarily amount to a rejection of such procedures. In fact, good engineering analysis depends critically on development of reliable models which can capture salient features of a process without employing unnecessary complexity. At the same time, analysts need to maintain proper caution regarding trustworthiness of answers produced with computer models. Many nonlinear systems respond strongly to small changes in physical parameters. Scientists today realize that, in dealing with highly nonlinear phenomena such as weather prediction, it is simply impossible to make reliable long term forecasts [38] because of various unalterable factors. Among these are a) uncertainty about initial conditions, b) uncertainty about the adequacy of mathematical models describing relevant physical processes, c) uncertainty about error contributions arising from use of spatial and time discretizations in construction of approximate numerical solutions, and d) uncertainty about effects of arithmetic roundoff error. In light of the criticism and cautions being stated about the dangers of using numerical solutions, the thrust of the discussion is that idealized models must not be regarded as infallible, and no numerical solution should be accepted as credible without adequately investigating effects of parameter perturbation within uncertainty limits of the parameters. To illustrate how sensitive a system can be to initial conditions, we might consider a very simple model concerning motion of a pendulum of length ℓ given an initial velocity v_0 starting from a vertically downward position. If v_0 exceeds $2\sqrt{g\ell}$, the pendulum will reach a vertically upward position and will go over the top. If v_0 is less than $2\sqrt{g\ell}$, the vertically upward position is never reached. Instead, the pendulum oscillates about the bottom position. Consequently, initial velocities of $1.999\sqrt{g\ell}$ and $2.001\sqrt{g\ell}$ produce quite different system behavior with only a tiny change in initial velocity. Other examples illustrating the difficulties of computing the response of nonlinear systems are cited below. These examples are not chosen to discourage use of the powerful tools now available for numerical integration of differential equations. Instead, the intent is to encourage users of these methods to exercise proper caution so that confidence in the reliability of results is fully justified.

Many important physical processes are governed by differential equations. Typical cases include dynamics of rigid and flexible bodies, heat conduction, and electrical current flow. Solving a system of differential equations subject to known initial conditions allows us to predict the future behavior of the related physical system. Since very few important differential equations can be solved in closed form, approximations which are directly or indirectly founded on series expansion methods have been developed. The basic problem addressed is that of accurately computing $Y(t+h)$ when $Y(t)$ is known, along with a differential equation governing system behavior from time t to $(t+h)$. Recursive application of a satisfactory numerical approximation procedure, with possible adjustment of step-size to maintain accuracy and stability, allows approximate prediction of system response subsequent to the starting time.

Numerical methods for solving differential equations are extremely important tools for analyzing engineering systems. Although valuable algorithms have been developed which facilitate construction of approximate solutions, all available methods are vulnerable to limitations inherent in the underlying approximation processes. The essence of the difficulty lies in the fact that, as long as a finite integration step-size is used, integration error occurs at each time step. In many instances, these errors have an accumulative effect which grows exponentially and eventually destroys solution validity. To some extent, accuracy problems can be limited by regulating step-size to keep local error within a desired tolerance. Typically, decreasing an integration tolerance increases the time span over which a numerical solution is valid. However, high costs for computer time to analyze large and complex systems sometimes preclude generation of long time histories which may be more expensive than is practically justifiable.

8.2   Runge-Kutta Methods and the ODE23 and ODE45 Integrators Provided in MATLAB

Formulation of one method to solve differential equations is discussed in this section. Suppose a function $y(x)$ satisfies a differential equation of the form $y'(x) = f(x, y)$, subject to $y(x_0) = y_0$, where f is a known differentiable function. We would like to compute an approximation of $y(x_0 + h)$ which agrees with a Taylor's series expansion up to a certain order of error. Hence,

$$y(x_0 + h) = \tilde{y}(x_0, h) + O(h^{n+1})$$

where $O(h^{n+1})$ denotes a quantity which decreases at least as fast as h^{n+1} for small h. Taylor's theorem allows us to write

$$
\begin{aligned}
y(x_0 + h) &= y(x_0) + y'(x_0)h + \frac{1}{2}y''(x_0)h^2 + O(h^3) \\
&= y_0 + f(x_0, y_0)h + \frac{1}{2}[f_x(x_0, y_0) + \\
&\quad f_y(x_0, y_0)f_0]h^2 + O(h^3)
\end{aligned}
$$

where $f_0 = f(x_0, y_0)$. The last formula can be used to compute a second order approximation $\hat{y}(x_0 + h)$, provided the partial derivatives f_x and f_y can be evaluated. However, this may be quite difficult since the function $f(x, y)$ may not even be known explicitly.

The idea leading to Runge-Kutta integration is to compute $y(x_0 + h)$ by making several evaluations of function f instead of having to differentiate that function. Let us seek an approximation in the form

$$\tilde{y}(x_0 + h) = y_0 + h[k_0 f_0 + k_1 f(x_0 + \alpha h, y_0 + \beta h f_0)]$$

We choose k_0, k_1, α, and β to make $\tilde{y}(x_0 + h)$ match the series expansion of $y(x)$ as well as possible. Evidently,

$$f(x_0 + \alpha h, y_0 + \beta h f_0) = f_0 + [f_x(x_0, y_0)\alpha + f_y(x_0, y_0)f_0\beta]h + O(h^2)$$

and therefore,

$$\begin{aligned}\tilde{y}(x_0 + h) &= y_0 + h[(k_0 + k_1)f_0 + k_1 \langle f_x(x_0, y_0)\alpha + \\ &\quad f_y(x_0, y_0)\beta f_0 \rangle]h + O(h^2) \\ &= y_0 + (k_0 + k_1)f_0 h + [f_x(x_0, y_0)\alpha k_1 + \\ &\quad f_y(x_0, y_0)f_0 \beta k_1]h^2 + O(h^3)\end{aligned}$$

The last relation shows that

$$y(x_0 + h) = \tilde{y}(x_0 + h) + O(h^3)$$

provided

$$k_0 + k_1 = 1 \qquad \alpha k_1 = \frac{1}{2} \qquad \beta k_1 = \frac{1}{2}$$

This system of three equations in four unknowns has an infinite number of solutions, one of these is $k_0 = k_1 = \frac{1}{2}$, $\alpha = \beta = 1$. This implies that

$$y(x_0 + h) = y(x_0) + \frac{1}{2}[f_0 + f(x_0 + h, y_0 + h f_0)]h + O(h^3)$$

Neglecting the truncation error $O(h^3)$ gives a difference approximation known as Heun's method [51], which is classified as a second order Runge-Kutta method. Reducing the step-size by h, reduces the truncation error by about a factor of $(\frac{1}{2})^3 = \frac{1}{8}$. Of course, the formula can be used recursively to compute approximations to $y(x_0 + h)$, $y(x_0 + 2h)$, $y(x_0 + 3h)$, ... In most instances, the solution accuracy decreases as the number of integration steps is increased and results eventually become unreliable. Decreasing h and taking more steps within a fixed time span helps, but this also has practical limits governed by computational time and arithmetic roundoff error.

The idea leading to Heun's method can be extended further to develop higher order formulas. One of the best known is the fourth order Runge-Kutta method described as follows

$$y(x_0 + h) = y(x_0) + h[k_1 + 2k_2 + 2k_3 + k_4]/6$$

where

$$k_1 = f(x_0, y_0) \quad k_2 = f(x_0 + \frac{h}{2}, y_0 + k_1\frac{h}{2})$$

$$k_3 = f(x_0 + \frac{h}{2}, y_0 + k_2\frac{h}{2}) \quad k_4 = f(x_0 + h, y_0 + k_3h)$$

The truncation error for this formula is order h^5, so the error is reduced by about a factor of $\frac{1}{32}$ when the step-size is halved. The development of the fourth order Runge-Kutta method is algebraically quite complicated [37]. We note that accuracy of order four is achieved with four evaluations of f for each integration step. This situation does not extend to higher orders. For instance, an eighth order formula may require twelve evaluations per step. This price of more function evaluations may be worthwhile provided the resulting truncation error is small enough to permit much larger integration steps than could be achieved with formulas of lower order. MATLAB provides function **ode45** which uses variable step-size and employs formulas of order four and five.

8.3   Step-size Limits Necessary to Maintain Numerical Stability

It can be shown that, for many numerical integration methods, taking too large a step-size produces absurdly large results which increase exponentially with successive time steps. This phenomenon, known as numerical instability, can be illustrated with the simple differential equation

$$y'(t) = f(t, y) = \lambda y$$

which has the solution $y = ce^{\lambda t}$. If the real part of λ is positive, the solution becomes unbounded with increasing time. However, a pure imaginary λ produces a bounded oscillatory solution, whereas the solution decays exponentially for $\textbf{real}(\lambda) < 0$. Applying Heun's method [37] gives

$$y(t + h) = y(t)[1 + (\lambda h) + \frac{(\lambda h)^2}{2}]$$

This shows that at each integration step the next value of y is obtained by multiplying the previous value by a factor

$$p = 1 + (\lambda h) + \frac{(\lambda h)^2}{2}$$

which agrees with the first three Taylor series terms of $e^{\lambda h}$. Clearly, the difference relation leads to

$$y_n = y_0 p^n$$

As n increases, y_n will approach infinity unless $|p| \le 1$. This stability condition can be interpreted geometrically by regarding λh as a complex variable z and solving for all values of z such that

$$1 + z + \frac{z^2}{2} = \zeta e^{\imath \theta} \qquad \zeta \le 1 \qquad 0 \le \theta \le 2\pi$$

Taking $\zeta = 1$ identifies the boundary of the stability region, which is normally a closed curve lying in the left half of the complex plane. Of course, h is assumed to be positive and the real part of λ is nonpositive. Otherwise, even the exact solution would grow exponentially. For a given λ, the step-size h must be taken small enough to make $|\lambda h|$ lie within the stability zone. The larger $|\lambda|$ is, the smaller h must be to prevent numerical instability.

The idea illustrated by Heun's method can be easily extended to a Runge-Kutta method of arbitrary order. A Runge-Kutta method of order n reproduces the exact solution through terms of order n in the Taylor series expansion. The differential equation $y' = \lambda y$ implies

$$y(t + h) = y(t)e^{\lambda h}$$

and

$$e^{\lambda h} = \sum_{k=0}^{n} \frac{(\lambda h)^k}{k!} + O(h^{n+1})$$

Consequently, points on the boundary of the stability region for a Runge-Kutta method of order n are found by solving the polynomial

$$1 - e^{i\theta} + \sum_{k=1}^{n} \frac{z^k}{k!} = 0$$

for a dense set of θ-values ranging from zero to 2π. Using MATLAB's intrinsic function **roots** allows easy calculation of the polynomial roots which may be plotted to show the stability boundary. The following short program accomplishes the task. Program output for integrators of order four and six are shown in Figures 8.1 and 8.2. Note that the region for order 4 resembles a semicircle with radius close to 2.8. Using $|\lambda h| > 2.8$, with Runge-Kutta of order 4, would give results which rapidly become unstable. The figures also show that the stability region for Runge-Kutta of order 6 extends farther out on the negative real axis than Runge-Kutta of order 4 does. The root finding process also introduces some meaningless stability zones in the right half plane which should be ignored.

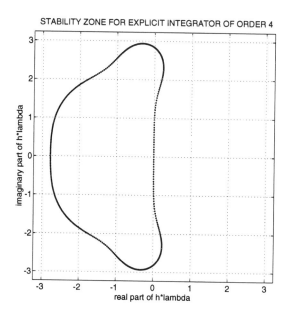

FIGURE 8.1. Stability Zone for Explicit Integrator of Order 4

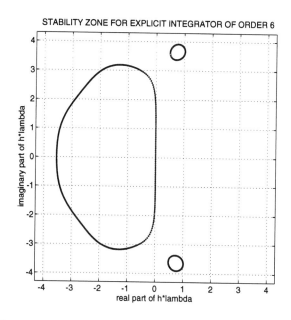

FIGURE 8.2. Stability Zone for Explicit Integrator of Order 6

MATLAB EXAMPLE

8.3.0.1 Script File rkdestab

```
1: % Example:  rkdestab
2: % ~~~~~~~~~~~~~~~~~~~~
3: % This program plots the boundary of the region
4: % of the complex plane governing the maximum
5: % step size which may be used for stability of
6: % a Runge-Kutta integrator of arbitrary order.
7: %
8: % npts  - a value determining the number of
9: %           points computed on the stability
10: %           boundary of an explicit Runge-Kutta
11: %           integrator.
12: % xrang - controls the square window within
13: %           which the diagram is drawn.
14: %           [ -3, 3, -3, 3] is appropriate for
15: %           the fourth order integrator.
16: %
17: % User m functions required: genprint
18: %------------------------------------------------
19:
20: clf;
21: fprintf('\nSTABILITY REGION FOR AN ')
22: fprintf('EXPLICIT RUNGE-KUTTA')
23: fprintf('\n    INTEGRATOR OF ARBITRARY ')
24: fprintf('ORDER\n')
25: nordr=input('Give the integrator order > ');
26: fprintf('\nInput the number of points ')
27: fprintf('used to define')
28: npts=input('the boundary (100 is typical) > ');
29: r=zeros(npts,nordr); v=1./gamma(nordr+1:-1:2);
30: d=2*pi/(npts-1); i=sqrt(-1);
31:
32: % Generate polynomial roots to define the
33: % stability boundary
34: for j=1:npts
35:   % polynomial coefficients
36:   v(nordr+1)=1-exp(i*(j-1)*d);
37:   % complex roots
38:   t=roots(v); r(j,:)=t(:).';
39: end
40:
```

```
41: % Plot the boundary
42: rel=real(r(:)); img=imag(r(:));
43: w=1.1*max(abs([rel;img]));
44: plot(rel,img,'w.');
45: axis([-w,w,-w,w]); axis('square')
46: xlabel('real part of h*lambda')
47: ylabel('imaginary part of h*lambda');
48: ns=int2str(nordr);
49: st=['STABILITY ZONE FOR EXPLICIT ' ...
50:     'INTEGRATOR OF ORDER ',ns];
51: title(st); grid; genprint('rkdestab.ps')
52:
53: disp(' '); disp('All Done')
```

8.4   Discussion of Procedures to Maintain Accuracy by Varying Integration Step-size

When we solve a differential equation numerically, our first inclination is to seek output at even increments of the independent variable. However, this is not the most natural form of output appropriate to maintain integration accuracy. Whenever solution components are changing rapidly, a small time step may be needed, whereas using a small time step might be quite inefficient at times where the solution remains smooth. Most modern ODE programs employ variable step-size algorithms which decrease the integration step-size whenever some local error tolerance is violated and conversely increase the step-size when the increase can be performed without loss of accuracy. If results at even time increments are needed, these can be determined by interpolation of the non-equidistant values.

Although the derivation of algorithms to regulate step-size is an important topic, development of these methods is not presented here. Several references [37, 39, 44, 51] discuss this topic with adequate detail. The primary objective in regulating step-size is to gain computational efficiency by taking as large a step-size as possible while maintaining accuracy and minimizing the number of function evaluations.

Practical problems involving a single first order differential equation are rarely encountered. More commonly, a system of second order equations occurs which is then transformed into a system involving twice as many first order equations. Several hundred, or even several thousand dependent variables may be involved. Evaluating the necessary time derivatives at a single time step may require computationally intensive tasks such as matrix inversion. Furthermore, performing this fundamental calculation several thousand times may be necessary in order to construct time responses over time intervals of practical interest. Integrating large systems of nonlinear differential equations is one of the most important and most resource intensive aspects of scientific computing.

Instead of deriving the algorithms used for step-size control in **ode45**, we will outline briefly the ideas employed to integrate $y'(t) = f(t, y)$ from t to $(t+h)$. It is helpful to think of y as a vector.

For a given time step and y value, the program makes six evaluations of f. These values allow evaluation of two Runge-Kutta
formulas, each having different truncation errors. These formulas
permit estimation of the actual truncation error and proper step-
size adjustment to control accuracy. If the estimated error is too
large, the step-size is decreased until the error tolerance is satisfied
or an error condition occurs because the necessary step-size has
fallen below a set limit. If the estimated error is found to be smaller
than necessary, the integration result is accepted and the step-size
is increased for the next pass. Even though this type of process
may not be extremely interesting to discuss, it is nevertheless a
completely essential part of any well designed program for integrating differential equations numerically. Readers should become
familiar with the error control feature of ODE solvers they employ. Printing and studying the code for **ode45** is worthwhile. It
should also be remembered that solutions generated by tools such
as **ode45** always contain accumulated error effects from roundoff and arithmetic truncation. Such errors eventually make results
sufficiently far from the starting time of the solution invalid.

This chapter concludes with the analysis of two realistic nonlinear problems having certain properties of their exact solutions
known. These known properties are compared with numerical results to assess error growth. The first problem studies an inverted
pendulum for which the loading function produces a simple exact
displacement function. The second problem treats the dynamics
of a falling chain constrained at both ends.

8.5 Example on Forced Oscillations of an Inverted Pendulum

The inverted pendulum in Figure 8.3 involves a weightless rigid
rod of length l which has a mass m attached to the end. Attached
to the mass is a spring with stiffness constant k and an unstretched
length of γl. The spring has length l when the pendulum is in
the vertical position. Externally applied loads consist of a driving
moment $M(t)$, the particle weight, and a viscous damping moment
$cl^2\dot{\theta}$.

The differential equation governing the motion of this system

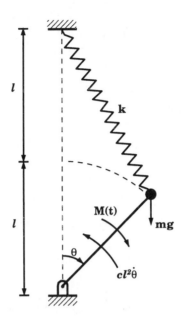

FIGURE 8.3. Forced Vibration of an Inverted Pendulum

is found to be

$$\ddot{\theta} = -(c/m)\dot{\theta} + (g/l)\sin(\theta) + M(t)/(ml^2) - (2k/m)\sin(\theta)(1 - \alpha/\lambda)$$

where

$$\lambda = \sqrt{5 - 4\cos(\theta)}$$

This system can be changed to a more convenient form by introducing dimensionless variables. We let $t = (\sqrt{l/g})\tau$ where τ is dimensionless time. Then

$$\ddot{\theta} = -\alpha\dot{\theta} + \sin(\theta) + P(\tau) - \beta\sin(\theta)(1 - \gamma/\lambda)$$

where

$$
\begin{aligned}
\alpha &= (c/m)\sqrt{l/g} = \text{viscous damping factor} \\
\beta &= 2(k/m)/(g/l) \\
\lambda &= \sqrt{5 - 4\cos(\theta)} \\
\gamma &= (\text{unstretched spring length})/l \\
P(\tau) &= M/(mgl) = \text{dimensionless driving moment}
\end{aligned}
$$

It is interesting to test how well a numerical method can reconstruct a known exact solution for a nonlinear function. Let us assume that the driving moment $M(\tau)$ produces a motion having the equation

$$\theta_e(\tau) = \theta_0 \sin(\omega \tau)$$

for arbitrary θ_0 and ω. Then

$$\dot{\theta}_e(\tau) = \omega \theta_0 \cos(\omega \tau)$$

and

$$\ddot{\theta}_e(\tau) = -\omega^2 \theta_e$$

Consequently, the necessary driving moment is

$$P(\tau) = -\omega^2 \theta_e - \sin(\theta_e) + \gamma \omega \theta_0 \cos(\omega \tau) +$$

$$\beta \sin(\theta_e) \left[1 - \gamma / \sqrt{5 - 4\cos(\theta_e)} \right]$$

Applying this forcing function, along with the initial conditions

$$\theta(0) = 0 \qquad \dot{\theta}(0) = \theta_0 \omega$$

should return the solution $\theta = \theta_e(\tau)$. For a specific numerical example we choose $\theta_0 = \pi/8$, $\omega = 0.5$, and four different combinations of β, γ, and tol. The second order differential equation has the form $\ddot{\theta} = f(\tau, \theta, \dot{\theta})$. This is expressed as a first order matrix system by letting $y_1 = \theta$, $y_2 = \dot{\theta}$; which gives

$$\dot{y}_1 = y_2 \qquad \dot{y}_2 = f(\tau, y_1, y_2)$$

A function describing the system for solution by **ode45** is provided at the end of this section. Parameters θ_0, ω_0, α, ζ, and β are passed as global variables.

We can examine how well the numerically integrated θ match θ_e by using the error measure

$$|\theta(\tau) - \theta_e(\tau)|$$

Furthermore, the exact solution satisfies

$$\theta_e^2 + (\dot{\theta}_e/\omega)^2 = \theta_0^2$$

so plotting $\dot{\theta}/(\theta_0 \omega)$ on a horizontal axis and θ/θ_0 on a vertical axis should produce a unit circle. Violation of that condition signals loss of solution accuracy.

How certain physical parameters and numerical tolerances affect terms in this problem can be demonstrated by the following four data cases.

1. The spring is soft and initially unstretched. A liberal integration tolerance is used.

2. The spring is soft and initially unstretched. A stringent integration tolerance is used.

3. The spring is stiff and initially stretched. A liberal integration tolerance is used.

4. The spring is stiff and initially stretched. A stringent integration tolerance is used.

The curves in Figure 8.4 show the following facts:

1. When the spring is unstretched initially, the numerical solution goes unstable quickly.

2. Stretching the spring initially and increasing the spring constant improves numerical stability of the solution.

3. Decreasing the integration tolerance increases the time period over which the solution is valid.

An additional curve illustrating the numerical inaccuracy of results for Case 1 appears in Figure 8.4. A plot of $\theta(\tau)$ versus $\dot{\theta}(\tau)/\omega$ should produce a circle. However, solution points quickly depart from the desired locus.

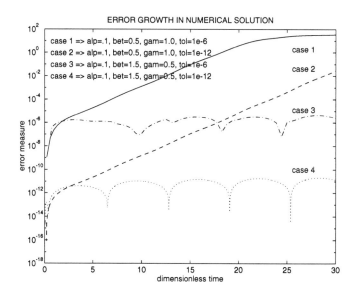

FIGURE 8.4. Error Growth in Numerical Solution

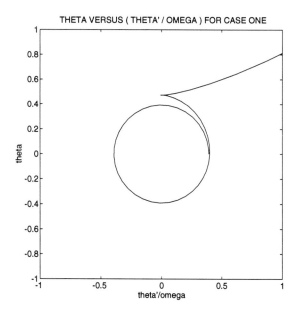

FIGURE 8.5. Case 1

MATLAB EXAMPLE

8.5.0.2   Script File prun

```
1:  % Example: prun
2:  % ~~~~~~~~~~~~~~
3:  % Dynamics of an inverted pendulum integrated
4:  % by use of ode45.
5:  %
6:  % User m functions required:
7:  %     pinvert, mom, genprint
8:  %-----------------------------------------------
9:
10: global th0_ w_ alp_ bet_ gam_ ncal_
11: th0_=pi/8; w_=.5; tmax=30; ncal_=0;
12:
13: fprintf('\nFORCED OSCILLATION OF AN ')
14: fprintf('INVERTED PENDULUM\n')
15: fprintf('\nNote: Several minutes are ')
16: fprintf('required to generate')
17: fprintf('\nthe four sets of numerical ')
18: fprintf('results needed.\n')
19:
20: % loose spring with liberal tolerance
21: alp_=.1; bet_=1.; gam_=1.; tol=1.e-6;
22: [t1,z1]= ...
23:    ode45('pinvert',0,tmax,[0;w_*th0_],tol,0);
24: n1=ncal_; ncal_=0;
25:
26: % loose spring with stringent tolerance
27: gam_=1.; bet_=1.;tol=1.e-12;
28: [t2,z2]= ...
29:    ode45('pinvert',0,tmax,[0;w_*th0_],tol,0);
30: n2=ncal_; ncal_=0;
31:
32: % tight spring with liberal tolerance
33: gam_=0.5; bet_=2.0; tol=1.e-6;
34: [t3,z3]= ...
35:    ode45('pinvert',0,tmax,[0;w_*th0_],tol,0);
36: n3=ncal_; ncal_=0;
37:
38: % tight spring with stringent tolerance
39: gam_=0.5; bet_=2.0; tol=1.e-12;
40: [t4,z4]= ...
```

```
41:    ode45('pinvert',0,tmax,[0;w_*th0_],tol,0);
42: n4=ncal_; ncal_=0; save pinvert.mat
43:
44: % Plot results
45: clf; semilogy( ...
46:    t1,abs(z1(:,1)/th0_-sin(w_*t1)),'-w',...
47:    t2,abs(z2(:,1)/th0_-sin(w_*t2)),'--w',...
48:    t3,abs(z3(:,1)/th0_-sin(w_*t3)),'-.w',...
49:    t4,abs(z4(:,1)/th0_-sin(w_*t4)),':w');
50: title('ERROR GROWTH IN NUMERICAL SOLUTION')
51: xlabel('dimensionless time');
52: ylabel('error measure');
53: fprintf('\nUse mouse to select the 8 ')
54: fprintf('legend locations');
55: fprintf('\n(1 - x,y for case 1)');
56: fprintf('\n(2 -   y for case 2)');
57: fprintf('\n(3 -   y for case 3)');
58: fprintf('\n(4 -   y for case 4)');
59: fprintf('\n(5 - x,y for case 1 => ...)');
60: fprintf('\n(6 -   y for case 2 => ...)');
61: fprintf('\n(7 -   y for case 3 => ...)');
62: fprintf('\n(8 -   y for case 4 => ...)\n');
63: [x,y]=ginput(8);
64: text(x(1),y(1),'case 1');
65: text(x(1),y(2),'case 2');
66: text(x(1),y(3),'case 3');
67: text(x(1),y(4),'case 4');
68: text(x(5),y(5),['case 1 => alp=.1, ' ...
69:      'bet=0.5, gam=1.0, tol=1e-6'])
70: text(x(5),y(6),['case 2 => alp=.1, ' ...
71:      'bet=0.5, gam=1.0, tol=1e-12'])
72: text(x(5),y(7),['case 3 => alp=.1, ' ...
73:      'bet=1.5, gam=0.5, tol=1e-6'])
74: text(x(5),y(8),['case 4 => alp=.1, ' ...
75:      'bet=1.5, gam=0.5, tol=1e-12'])
76: fprintf('\nPress Return to continue\n'); pause;
77: genprint('pinvert.ps')
78:
79: % plot a phase diagram for case 1
80: plot(z1(:,2)/w_,z1(:,1));
81: axis('square'); axis([-1,1,-1,1]);
82: xlabel('theta''/omega'); ylabel('theta')
83: title(['THETA VERSUS ( THETA'' / OMEGA ) ' ...
84:        'FOR CASE ONE'])
85: genprint('crclplt.ps')
```

```
86: disp(' '); disp('All Done')
```

8.5.0.3   Function pinvert

```
 1: function zdot=pinvert(t,z)
 2: %
 3: % zdot=pinvert(t,z)
 4: % ~~~~~~~~~~~~~~~~~
 5: % Equation of motion for the pendulum
 6: %
 7: % t    - time value
 8: % z    - vector [theta ; theta_dot]
 9: % zdot - time derivative of z
10: %
11: % User m functions called:  mom
12: %-------------------------------------------------
13:
14: global alp_ bet_ gam_ ncal_
15: ncal_=ncal_+1; th=z(1); thd=z(2);
16: c=cos(th); s=sin(th); lam=sqrt(5-4*c);
17: zdot=[thd;
18:       mom(t)+s-alp_*thd-bet_*s*(1-gam_/lam)];
```

8.5.0.4   Function mom

```
 1: function me=mom(t)
 2: %
 3: % me=mom(t)
 4: % ~~~~~~~~~
 5: % t  - time
 6: % me - driving moment needed to produce
 7: %      exact solution
 8: %
 9: % User m functions called:  none.
10: %-----------------------------------------------
11:
12: global th0_ w_ alp_ bet_ gam_
13: th=th0_*sin(w_*t);
14: thd=w_*th0_*cos(w_*t); thdd=-th*w_^2;
15: s=sin(th); c=cos(th); lam=sqrt(5-4*c);
16: me=thdd-s+alp_*thd+bet_*s*(1-gam_/lam);
```

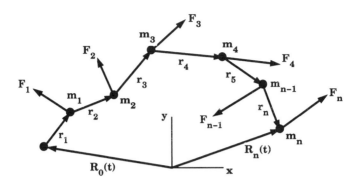

FIGURE 8.6. Chain with Specified End Motion

8.6 Example on Dynamics of a Chain with Specified End Motion

The dynamics of flexible cables is often modeled using a chain of rigid links connected by frictionless joints. A chain having specified end motions illustrates the behavior of a system governed by nonlinear equations of motion and auxiliary algebraic constraints. In particular, we will study a gravity loaded cable fixed at both ends. The total cable length exceeds the distance between supports, so that the static deflection configuration would resemble a catenary.

A simple derivation of the equations of motion employing principles of rigid body dynamics is given next. Readers not versed in principles of rigid body dynamics [41] may nevertheless understand the subsequent programs by analyzing the equations of motion which have a concise mathematical form. The numerical solutions vividly illustrate some numerical difficulties typically encountered in multibody dynamical studies. Such problems are both computationally intensive, as well as highly sensitive to accumulated effects of numerical error.

The mathematical model of interest is the two-dimensional motion of a cable (or chain) having n rigid links connected by frictionless joints. A typical link i has its mass m_i concentrated at one end. The geometry is depicted in Figure 8.6. The chain ends undergo specified motions $R_0(t) = [X_0(t) \; ; \; Y_0(t)]$ for the first link and $R_n(t) = [X_n(t) ; Y_n(t)]$ for the last link. The direction vector along link $\imath$ is described by $r_\imath = [x_\imath ; y_\imath] = \ell_\imath[\cos(\theta_\imath) ; \sin(\theta_\imath)]$. We assume that each joint $\imath$ is subjected to a force $F_\imath = [f_{x\imath} ; f_{y\imath}]$ where $0 \le \imath \le n$. Index values $\imath = 0$ and $\imath = n$ denote unknown constraint forces which must act at the outer ends of the first and last links to achieve the required end displacements. The forces applied at the interior joints are arbitrary. It is convenient to characterize the dynamics of each link in terms of its direction angle. Thus

$$\dot{r}_\imath = r'_\imath \dot{\theta}_\imath \qquad \ddot{r}_\imath = r'_\imath \ddot{\theta}_\imath + r''_\imath \dot{\theta}_\imath^2 = r'_\imath \ddot{\theta}_\imath - r_\imath \dot{\theta}_\imath^2$$

where primes and dots denote differentiation with respect to $\theta_\imath$ and t, respectively. Therefore

$$\dot{r}_\imath = [-y_\imath ; x_\imath]\dot{\theta}_\imath \qquad \ddot{r}_\imath = [-y_\imath ; x_\imath]\ddot{\theta}_\imath - [x_\imath ; y_\imath]\dot{\theta}_\imath^2$$

The global position vector of joint $\imath$ is

$$R_\imath = R_0 + \sum_{j=1}^{\imath} r_j = R_0 + \sum_{j=1}^{n} < \imath - j > r_j$$

where the symbol $< k >$ means one for $k \ge 0$, and zero for negative k. Consequently, the velocity and acceleration of joint $\imath$ are

$$\dot{R}_\imath = \dot{R}_0 + \sum_{j=1}^{n} < \imath - j > r'_j \dot{\theta}_j$$

$$\ddot{R}_\imath = \ddot{R}_0 + \sum_{j=1}^{n} < \imath - j > r'_j \ddot{\theta}_j - \sum_{j=1}^{n} < \imath - j > r_j \dot{\theta}_j^2$$

The ends of the chain each have specified motions, so not all of the inclination angles are independent and consequently

$$\sum_{j=1}^{n} r_j = R_n - R_0$$

$$\sum_{j=1}^{n} r'_j \dot{\theta}_j = \dot{R}_n - \dot{R}_0$$

$$\sum_{j=1}^{n} r'_j \ddot{\theta}_j - \sum_{j=1}^{n} r_j \dot{\theta}_j^2 = \ddot{R}_n - \ddot{R}_0$$

Combining the last constraint equation with equations of motion written for masses $m_1, \cdots, m_n$ yields a complete system of $(n+2)$ equations determining $\ddot{\theta}_1, \cdots, \ddot{\theta}_n$ and the components of F_n. The fact that all masses are concentrated at frictionless joints shows that link i is a two-force member carrying an internal load directed along r_i. Consequently, the d'Alembert principle [41] implies that the sum of all external and inertial loads from joints $i, i+1, \cdots, n$ must give a resultant passing through joint i in the direction of r_i. Since r'_i and r_i are perpendicular, requiring a vector to be in the direction of r_i is equivalent to making it normal to r'_i. Therefore

$$r'_i \cdot \left[\sum_{j=1}^{n} <j-i> \left\{ F_j - m_j \ddot{R}_j \right\} \right] = 0 \qquad 1 \le i \le n$$

The last n equations involve $\ddot{\theta}_i$ and two end force components f_{xn} and f_{yn}. Some algebraic rearrangement results in a matrix differential equation of concise form containing several auxiliary coefficients defined as follows:

$$b_i = \sum_{k=i}^{n} m_k \qquad m_{ij} = m_{ij} = b_i \qquad 1 \le i \le n \qquad 1 \le j \le i$$

$$a_{ij} = m_{ij}(x_i x_j + y_i y_j) \qquad 1 \le i \le n \qquad 1 \le j \le n$$

$$b_{ij} = m_{ij}(x_i y_j - x_j y_i) \qquad 1 \le i \le n \qquad 1 \le j \le n$$

$$p_{xi} = \sum_{j=i}^{n-1} f_{xi} \qquad p_{yi} = \sum_{j=i}^{n-1} f_{yi} \qquad 1 \le i \le n$$

For $i = n$, the last two sums mean $p_{xn} = p_{yn} = 0$. Furthermore, we denote the acceleration components of the chain ends as $\ddot{R}_0 = [a_{xo} \; ; \; a_{y0}]$ and $\ddot{R}_n = [a_{xn} \; ; \; a_{yn}]$. Using the various quantities just defined, the equations of motion become

$$\sum_{j=1}^{n} a_{ij} \ddot{\theta}_j + y_i f_{xn} - x_i f_{yn} = \sum_{j=1}^{n} b_{ij} \dot{\theta}_j^2 + x_i(p_{yi} - b_i a_{y0}) -$$

$$y_i(p_{xi} - b_i a_{x0})$$

$$= e_i \qquad 1 \le i \le n$$

The remaining two components of the constraint equations completing the system are

$$\sum_{j=1}^{n} y_j \ddot{\theta}_j = -\sum_{j=1}^{n} x_j \dot{\theta}_j^2 - a_{xn} + a_{x0} = e_{n+1}$$

$$\sum_{j=1}^{n} x_j \ddot{\theta}_j = \sum_{j=1}^{n} y_j \dot{\theta}_j^2 + a_{yn} - a_{y0} = e_{n+2}$$

Consequently, we get the following symmetric matrix equation to solve for $\ddot{\theta}_1, \cdots, \ddot{\theta}_n, f_{xn}$ and f_{yn}.

$$\begin{bmatrix} A & X & Y \\ X^T & 0 & 0 \\ Y^T & 0 & 0 \end{bmatrix} \begin{bmatrix} \ddot{\theta} \\ f_{xn} \\ -f_{yn} \end{bmatrix} = \begin{bmatrix} E \end{bmatrix}$$

where X, Y, E and θ are column matrices, and the matrix $A = [a_{ij}]$ is symmetric. Because most numerical integrators for differential equations solve first order systems, it is convenient to employ the vector $Z = [\theta \; ; \; \dot{\theta}]$ having $2n$ components. Then the differential equation $\dot{Z} = H(t, Z)$ is completely defined when $\ddot{\theta}$ has been computed for known Z. The system is integrated numerically to give θ and $\dot{\theta}$ as functions of time. These quantities can then be used to compute the global cartesian coordinates of the link configurations, thereby completely describing the time history of the chain.

The general equations of motion simplify somewhat when the chain ends are fixed and the external forces only involve gravity loads. Then $p_{xi} = 0$ and $p_{yi} = -g(b_i - b_n)$ which gives

$$\sum_{j=1}^{n} m_{ij}(x_i x_j + y_i y_j)\ddot{\theta}_j - x_i f_{yn} + y_i f_{xn} =$$

$$g(b_i - b_n) + \sum_{j=1}^{n} m_{ij}(x_i y_j - x_j y_i)\dot{\theta}_j^2 \qquad 1 \le i \le n$$

The last two equations to complete the set are:

$$\sum_{j=1}^{n} x_j \ddot{\theta}_j = \sum_{j=1}^{n} y_j \dot{\theta}_j^2 \qquad \sum_{j=1}^{n} y_j \ddot{\theta}_j = -\sum_{j=1}^{n} x_j \dot{\theta}_j^2$$

A program was written to simulate motion of a cable fixed at both ends and released from rest. The cable falls under the influence of gravity from an initially elevated position. Function **ode45** is used to perform the numerical integration. The program involves the following modules.

cablenl	driver program to set initial physical constants and numerical tolerances
equamo	function which defines the equations of motion for use by **ode45**
pltxmidl	function to plot the horizontal position of the middle
unsymerr	function to compute and plot a measure of how much the deflection platform loses symmetry with passing time
plotmotn	function producing an animated plot of the cable position for a sequence of times
eventime	function which linearly interpolates **ode45** output to produce position values corresponding to equidistant time intervals
lintrp	function performing piecewise linear interpolation

A configuration with eight identical links was specified. For simplicity, the total mass, total cable length, and gravity constant were all normalized to equal unity. Results of the simulation are shown below. Figure 8.7 shows cable positions during the early stages of motion when results of the numerical integration are reliable. However, the numerical solution eventually becomes worthless due to accumulated numerical inaccuracies yielding the motion predictions indicated in Figure 8.8. The nature of the error growth can be seen clearly in Figure 8.9 which plots the x-coordinate of the chain midpoint as a function of time. Since the chosen mass distribution and initial deflection is symmetrical about the middle, the subsequent motion will remain symmetrical unless the numerical solution becomes invalid. Although the midpoint coordinate should remain at a constant value of $0.5\sqrt{2}$, it appears to abruptly go unstable near $t = 17$. More careful examination indicates that this numerical instability does not actually occur suddenly. Instead, it grows exponentially from the outset of the simulation. The error is caused by the accumulation of truncation errors intrinsic to the numerical integration process allowing errors at each step which are regulated within a small but finite tolerance. A global measure of symmetry loss of the deflection pattern is plotted on a semilog scale in Figure 8.10. Note that the error curve has a nearly linear slope until the solution degenerates completely near $t = 17$. The reader can verify that choosing a less stringent error tolerance such as tol = 1E-3 will produce a solution which degenerates much sooner than $t = 17$. It should also be observed that this dynamical model exhibits another important characteristic of highly nonlinear systems, namely, extreme sensitivity to physical properties. Note that shortening the last link by only one part in ten thousand makes the system deflection quickly lose all appearance of symmetry by $t = 6$. Hence, two systems having nearly identical physical parameters and initial conditions may behave very differently a short time after motion is initiated. The conclusion implied is that analysts should thoroughly explore how parameter variations affect response predictions in nonlinear models.

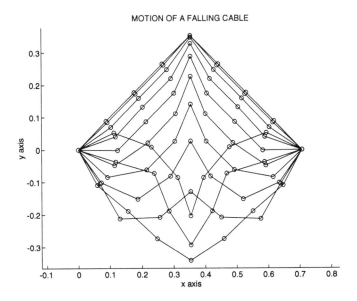

FIGURE 8.7. Motion During Initial Phase

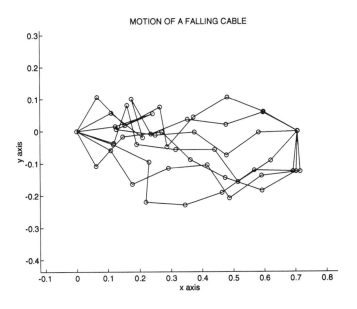

FIGURE 8.8. Motion After Solution Degenerates

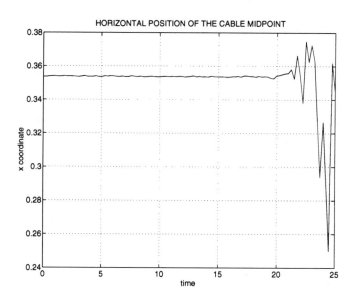

FIGURE 8.9. Horizontal Position of the Cable Midpoint

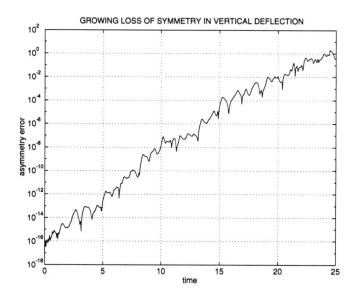

FIGURE 8.10. Growing Loss of Symmetry in Vertical Deflection

MATLAB EXAMPLE

8.6.0.5   Script File cablenl

```
 1: % Example: cablenl
 2: % ~~~~~~~~~~~~~~~~~
 3: % Numerical integration of the matrix
 4: % differential equations for the nonlinear
 5: % dynamics of a cable fixed at both ends.
 6: %
 7: % User m functions required:
 8: %     unsymerr, plotmotn, eventime, equamo,
 9: %     lintrp, removfil, pltxmidl, genprint
10: %-----------------------------------------------
11:
12: clear; clf;
13:
14: % Make variables global for use
15: % by other routines
16: global first_ n_ m_ len_ grav_ b_ mas_ py_
17:
18: fprintf('\nNONLINEAR DYNAMICS OF A ')
19: fprintf('FALLING CABLE\n')
20: fprintf('\nPlease wait: calculations will ')
21: fprintf('take several minutes\n');
22:
23: % Set up data for a cable of n_ links,
24: % initially arranged in a triangular
25: % deflection configuration.
26:
27: % parameter controlling initialization of
28: % auxiliary parameters used in function
29: % equamo
30: first_=1;
31: % number of links in the cable
32: n_=8; n=n_; nh=n_/2;
33: % vector of lengths and gravity constant
34: len_=1/n*ones(n,1); grav_=1;
35: % vector of mass constants
36: m_=ones(1,n_)/n_;
37:
38: % initial position angles
39: th0=pi/4*[ones(nh,1);-ones(nh,1)];
40: td0=zeros(size(th0)); z0=[th0;td0];
```

```
41:
42: % time limits, integration tolerances, and
43: % number of plot points
44: tfinl=25; tmin=0; tmax=25;
45: tol=1e-6; nplot=100;
46: trac=0; len=len_;
47:
48: % Perform the numerical integration using a
49: % variable stepsize Runge-Kutta integrator
50:
51: quadtime=cputime; flps=flops;
52: %............................................
53: % Use next line for M-file implementation
54: [tu,zu]=ode45('equamo',0,tfinl,z0,tol,trac);
55: %
56: % Use next 2 lines for Mex-file implementation
57: %[tu,zu,rerr,aerr,istat]= ...
58: %   ode('equamo',0,tfinl,z0,tol,tol,1000,1);
59: %............................................
60: thu=zu(:,1:n);
61: quadtime=cputime-quadtime; flps=flops-flps;
62:
63: % Interpolate for results evenly spaced in time
64: [tevn,xevn,yevn]= ...
65:     eventime(tu,thu,len,nplot,tmin,tmax);
66: save cablenl.mat
67:
68: % Plot the horizontal position of the
69: % cable midpoint
70: pltxmidl(tevn,xevn(:,1+n_ /2));
71: fprintf('\nPress RETURN to continue\n'); pause
72:
73: % Show error growth indicated by symmetry
74: % loss of the vertical deflection pattern
75: [terr,yerr]=unsymerr(tu,thu,len,tmin,tmax);
76: plotmotn(xevn,yevn)
77: clf; fprintf('\nAll Done\n')
```

8.6.0.6 Function equamo

```
1: function zdot=equamo(t,z)
2: %
3: % zdot=equamo(t,z)
```

```
 4: %  ~~~~~~~~~~~~~~~~
 5: % Equation of motion for a cable fixed at
 6: % both ends and loaded by gravity forces only
 7: %
 8: % t         current time value
 9: % z         column vector defined by
10: %           [thet(t);theta'(t)]
11: % zdot      column vector defined by
12: %           the concatenation
13: %           z'(t) = [theta'(t);theta''(t)]
14: %
15: % User m functions called:  none.
16: %---------------------------------------------
17:
18: % Values accessed as global variables
19: global first_ n_ m_ len_ grav_ b_ mas_ py_
20:
21: % Initialize parameters first time
22: % function is called
23: if first_==1, first_=0;
24: % mass parameters
25:   m_=m_(:); b_=flipud(cumsum(flipud(m_)));
26:   mas_=b_(:,ones(n_,1));
27:   mas_=tril(mas_)+tril(mas_,-1)';
28: % load effects from gravity forces
29:   py_=-grav_*(b_-b_(n_));
30: end
31:
32: % Solve for zdot = [theta'(t); theta''(t)];
33: n=n_; len=len_;
34: th=z(1:n); td=z(n+1:2*n); td2=td.*td;
35: x=len.*cos(th); y=len.*sin(th);
36:
37: % Matrix of mass coefficients and
38: % constraint conditions
39: amat=[[mas_.*(x*x'+y*y'),x,y];
40:      [x,y;zeros(2,2)]'];
41:
42: % Right side vector involves applied forces
43: % and inertial terms
44: bmat=x*y'; bmat=mas_.*(bmat-bmat');
45:
46: % Solve for angular acceleration.
47: % Most computation occurs here.
48: soln=amat\[x.*py_+bmat*td2; y'*td2; -x'*td2];
```

```
49:
50: % Final result needed for use by the numerical
51: % integrator
52: zdot=[td; soln(1:n)];
```

8.6.0.7   Function unsymerr

```
 1: function [terr,yerr]= ...
 2:                   unsymerr(t,thta,len,tmin,tmax)
 3: %
 4: % [terr,yerr]=unsymerr(t,thta,len,tmin,tmax)
 5: % ~~~~~~~~~~~~~~~~~~~~~~~~~~~~~~~~~~~~~~~~~~~~~
 6: % This function computes an error measure which
 7: % shows how the initally symmetric deflection
 8: % configuration progressively loses symmetry
 9: % due to growing error in the numerical
10: % solution.
11: %
12: % t         - vector of times for solution
13: %             evaluation
14: % thta      - matrix with successive rows
15: %             defining sets of theta values
16: %             which specify a configuration of
17: %             the cable
18: % len       - vector of lengths of the
19: %             cable links
20: % tmin,tmax - time limits over which solution
21: %             error is evaluated
22: % terr      - subset of t values between tmin
23: %             and tmax at which the solution
24: %             error is computed
25: % yerr      - vector specifying the solution
26: %             error defined in the following
27: %             manner. Let Y be a column vector
28: %             of length n-1 representing the
29: %             vertical deflection of the cable
30: %             (excluding the fixed ends). A
31: %             corresponding element of yerr
32: %             would be defined as
33: %             norm(Y-flipud(Y))/norm(Y).
34: %
35: % User m functions called:  genprint
36: %-------------------------------------------------
```

```
37:
38: [nt,n]=size(thta);
39: if nargin < 5, tmin=min(t); tmax=max(t); end
40:
41: % Compute time within specified limits
42: n1=sum(t<=tmin); n2=sum(t<=tmax);
43: terr=t(n1:n2); thta=thta(n1:n2,:);
44:
45: % Compute values of the vertical deflection.
46: nte=length(terr); len=len(:)';
47: y=cumsum((len(ones(nte,1),:).*sin(thta))')';
48:
49: % Evaluate growth in unsymmetrical character
50: % of the deflection pattern
51: yy=y(:,1:n-1); ydif=yy-yy(:,n-1:-1:1);
52: yerr=sqrt(sum((ydif.*ydif)'))./ ...
53:     sqrt(sum((yy.*yy)'));
54:
55: hold off; axis('normal'); clf
56:
57: % Graph the solution error. An approximately
58: % linear trend on a semilog plot would
59: % indicate exponential growth of error with
60: % passing time.
61: semilogy(terr,yerr);
62: xlabel('time'); ylabel('asymmetry error');
63: title(['GROWING LOSS OF SYMMETRY IN ' ...
64:     'VERTICAL DEFLECTION']);
65: grid
66: disp('Press RETURN to continue'); pause
67: genprint('unsymerr');
```

8.6.0.8 Function plotmotn

```
1: function plotmotn(x,y,isave)
2: %
3: % plotmotn(x,y,ifsave)
4: % ~~~~~~~~~~~~~~~~~~~~
5: % This function plots the cable time
6: % history described by coordinate values
7: % stored in the rows of matrices x and y.
8: %
9: % x,y   - matrices having successive rows
```

```
10: %              which describe position
11: %              configurations for the cable
12: % isave - parameter controlling the form
13: %              of output. When isave is not input,
14: %              successive positions are plotted.
15: %              The next position is plotted when
16: %              the user presses any key. If isave
17: %              is given a value, then successive
18: %              positions are not erased so that
19: %              the sequence of positions are all
20: %              left showing.
21: %
22: % User m functions called:  genprint
23: %------------------------------------------------
24:
25: % Set a square window to contain all
26: % possible positions
27: [nt,n]=size(x);
28: xmin=min(x(:)); xmax=max(x(:));
29: ymin=min(y(:)); ymax=max(y(:));
30: w=max(xmax-xmin,ymax-ymin)/2;
31: xmd=(xmin+xmax)/2; ymd=(ymin+ymax)/2;
32: w=1.05*w;
33: hold off; clf; axis('normal')
34: axis('equal');
35: axis([xmd-w,xmd+w,ymd-w,ymd+w]);
36: title('MOTION OF A FALLING CABLE')
37: xlabel('x axis'); ylabel('y axis')
38:
39: % Plot successive positions describing
40: % time history
41: istr=['Press RETURN for next position, ' ...
42:       'Q RETURN to exit'];
43: hold on
44: for j=1:nt
45:     fprintf('Position %6.0f of %6.0f\n', j, nt);
46:     xj=x(j,:); yj=y(j,:);
47:     if nargin==2
48:         % Plot and then erase
49:         plot(xj,yj,'-w');
50:         more=input(istr,'s');
51:         if more == 'Q' | more == 'q', break, end
52:         cla;
53:     else
54:         % Plot and leave trace
```

```
55:        plot(xj,yj,'-w'); plot(xj,yj,'ow');
56:        more=input(istr,'s');
57:        if more == 'Q' | more == 'q', break, end
58:    end
59: end
60:
61: % Save plot history for subsequent printing
62: genprint('plotmotn');
63:
64: hold off; axis;
```

8.6.0.9 Function eventime

```
 1: function [te,xe,ye]= ...
 2:                eventime(t,th,len,nte,tmin,tmax)
 3: %
 4: % [te,xe,ye]=eventime(t,th,len,nte,tmin,tmax)
 5: % ~~~~~~~~~~~~~~~~~~~~~~~~~~~~~~~~~~~~~~~~~~~~~
 6: % This function computes cable position
 7: % coordinates for a series of evenly spaced
 8: % time values
 9: %
10: % t          unevenly spaced time values
11: %            produced by the differential
12: %            equation integrator
13: % th         theta values output by the
14: %            differential equation integrator
15: % len        vector of lengths for the
16: %            cable links
17: % nte        number of evenly spaced time
18: %            values to be used
19: % tmin,tmax  maximum and minimum times
20: %            for output
21: % te         vector of output times
22: % xe,ye      matrices containing interpolated
23: %            coordinate values corresponding
24: %            to even time increments
25: %
26: % User m functions called:  lintrp
27: %-------------------------------------------------
28:
29: % Compute position vectors xe, ye corresponding
30: % to evenly spaced times from tmin to tmax
```

```
31: if nargin < 6, tmin=min(t); tmax=max(t); end
32: [nt,n]=size(th); len=len(:)';
33:
34: % Generate vector of equally spaced times
35: te=tmin+(tmax-tmin)/(nte-1)*(0:nte-1)';
36: the=zeros(nte,n);
37:
38: % Compute theta values at desired times
39: for j=1:n, the(:,j)=lintrp(te,t,th(:,j)); end
40:
41: % Generate global position coordinates
42: % for the desired times
43: xe=cumsum((len(ones(nte,1),:).*cos(the))')';
44: ye=cumsum((len(ones(nte,1),:).*sin(the))')';
45: xe=[zeros(nte,1),xe]; ye=[zeros(nte,1),ye];
```

8.6.0.10   Function pltxmidl

```
 1: function pltxmidl(t,x)
 2: %
 3: % pltxmidl(t,x)
 4: % ~~~~~~~~~~~~~
 5: % t - time vector
 6: % x - horizontal position of the cable midpoint
 7: %
 8: % User m functions called:  genprint
 9: %------------------------------------------------
10:
11: plot(t,x);
12: ylabel('x coordinate'); xlabel('time')
13: title(['HORIZONTAL POSITION OF THE ' ...
14:         'CABLE MIDPOINT'])
15: grid; genprint('xmidl.ps');
```

8.7 FORTRAN MEX Implementation: Dynamics of a Chain with Specified End Motion

8.7.1 INTRODUCTION

Although MATLAB includes a robust programming language and a comprehensive selection of intrinsic functions, occasionally there may be a desire to utilize a preexisting FORTRAN (or C) implementation of an algorithm. Users are encouraged to rewrite these algorithms in MATLAB's native language. However, for routines with several thousand lines of FORTRAN code, translating FORTRAN to MATLAB M-files may be impractical. MATLAB incorporates the capability to call FORTRAN (or C) routines from M-files and FORTRAN (or C) routines may call MATLAB's intrinsic routines and user written M-files. FORTRAN and C routines used in this fashion are referred to as MEX-files (MATLAB EXtensions).

In this section we will discuss and demonstrate the use of MEX-files by substituting a publicly-available FORTRAN ODE routine in place of the intrinsic routine **ode45** used in the example problem from the previous section (this routine is available via email request from the software collection at Oak Ridge National Laboratory). This problem is of interest because it typifies two important situations: 1) where MATLAB calls a FORTRAN routine, and 2) where a FORTRAN routine calls MATLAB. Figure 8.11 diagrams the general flow for the example problem MEX implementation. The MATLAB driver routine **cablenl** invokes the call to the FORTRAN MEX function **ode**. MATLAB intercepts this call and transfers program flow to the MATLAB-to-FORTRAN gateway routine **odeg**. This gateway routine then calls the actual FORTRAN integration routine **ode** which is a well known Adams type differential equation solver. Then **ode** must make its integrand calls by evaluation of the MATLAB code specified in **equamo**. To accomplish this, **ode** calls the FORTRAN-to-MATLAB gateway routine **odeext**. Finally, **odeext** invokes the M-file **equamo** to evaluate the function. The next section provides a detailed discussion of the issues related to a MEX implementation of the chain dynamics example.

Before discussing the specifics of the necessary code interfacing steps, we want to emphasize that the Adams integrator introduced here is not proposed as being superior to **ode45**. It simply typifies the code interfacing problem. Similar procedures could be used for other differential equation solvers.

8.7.2   MEX-ROUTINE DEVELOPMENT

The successful implementation of FORTRAN MEX-routines is dependent on a basic understanding of programming techniques not traditionally utilized by FORTRAN programmers. Additionally, the method which MATLAB employs to provide "hooks" to external routines must be understood.[1] Once these concepts are mastered, the reader should be able to develop MEX-routines independently.

8.7.2.1   Gateway Routine Structure and Implementation

All external MEX-routines interface with MATLAB using a standard method which will be illustrated by example. A MEX-routine is invoked via a function call from an M-file. Consider the M-file code extract below which invokes a FORTRAN MEX-routine by the name **fcall**.

```
    . . .
a=10; b=100.23; c=pi; d=ones(10,10);
[x,y,z]=fcall(a,b,c,d);
    . . .
```

MATLAB will first search to resolve the function reference by checking intrinsic functions, user M-files, and finally, user MEX-files. If the reference is determined to be a MEX-file, the following sequence occurs:

a) the number of arguments passed to the function are determined (This variable is called **NRHS** in MATLAB manuals[2] and is short for *number on right-hand side*. For this example

---

[1] The manual titled "External Interface Guide" contains a complete discussion of MATLAB's MEX capabilities and is included with the standard MATLAB distribution.

[2] The names NRHS, NLHS, PRHS, and PLHS are not reserved names and are used here for consistency only.

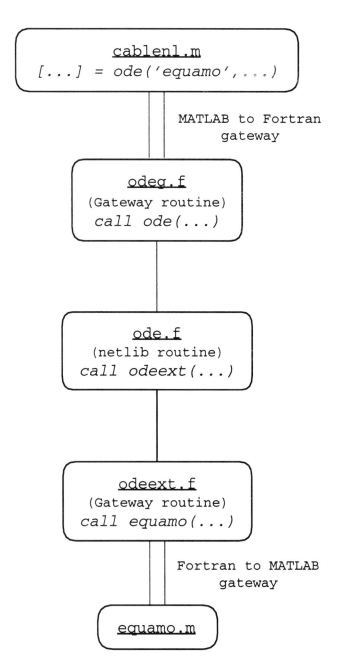

FIGURE 8.11. Flow Diagram for MEX-Routines

NRHS=4),

b) the number of arguments returned from the function are determined (This variable is called NLHS in MATLAB manuals and is short for *number on left-hand side*. For this example NLHS=3),

c) and two integer arrays are constructed which contain the integer addresses, or pointers, for each argument. (For the arguments being passed to the function this array is called PRHS. For the arguments returned from the function this array is called PLHS. In this example PRHS(1), PRHS(2), PRHS(3), and PRHS(4) would contain the integer address for a, b, c, and d, respectfully. Additionally, PLHS(1), PLHS(2), and PLHS(3) would contain, on return, the integer address for x, y, and z, respectfully.)

Several additional characteristics of passing arguments should also be noted.

- Arguments being passed to a function canNOT have their values changed. (MATLAB actually copies these variables to dummy storage prior to the call.)

- The length of the argument lists need not be consistent. (e.g. even though a function could have 5 arguments, the user can specify fewer in the call list. This is handy when default values are assumed unless explicitly defined.)

- Internally, MATLAB only understands a single type of data structure — a matrix. All quantities are stored as 64-bit floating point numbers.[3] Scalars and vectors are simply special classes of matrices. Matrices can have different characteristics (e.g. real/complex, number of rows, number of columns, etc.). The "pointer," or address, to a MATLAB matrix is actually a pointer to the beginning address of a set of memory locations containing the characteristics of the matrix and the matrix itself. This is called a data structure in languages such as C. (Each of the values contained in

---

[3]Strings are stored by placing the ASCII value for each character in a separate 64-bit location. Therefore, strings are quite wasteful of memory.

PRHS and PLHS represent the beginning address for the set of memory locations associated with a particular matrix.)

- Matrices are stored in column major order. This is the same as FORTRAN and opposite of C.

The standard distribution of MATLAB includes a set of FORTRAN (and C) service routines to facilitate the implementation of MEX-routines. These routines perform many of the cumbersome tasks associated with the data structure management and manipulation. The only issue which may be unfamiliar to FORTRAN programs is the manipulation of pointers. The remainder of this section will present a detailed discussion of the implementation specifics for the substitute ODE routine.

The implementation of a FORTRAN MEX-routine always requires the development of a "gateway" routine to serve as a bridge between MATLAB and the desired FORTRAN module. The gateway routine takes care of tasks such as: allocation of temporary matrices, checking argument list validity, assigning default variables, error trapping, calling the FORTRAN computation routine, and releasing allocated memory. A standard naming convention has been adopted for these gateway routines. For our example the computational routine is stored in a file named **ode.f**. Therefore, MATLAB expects the gateway routine to have the filename **odeg.f**.[4]

The actual M-file function call which invokes the gateway routine **odeg.f** for the example program is:

```
[t,x,rerr,aerr,istat]= ...
  ode('equamo',t0,tf,x0,relerr,abserr,ngrid,iflag)
```

The user should edit the M-file **cablenl**, lines 52-59, to replace the use of **ode45** with the MEX-routine **ode**. The table below provides a line-by-line synopsis of the gateway interface. The reader is advised to carefully study the code and the descriptions provided in the table to gain an appreciation for the subtleties of developing MEX gateway routines.

---

[4]Note that the .f filename extension follows the standard UNIX notation for FORTRAN routines.

odeg	
Line	Operation
109-110	**mexfunction** is standard "entry" point for gateway routines. Note the four parameters in the argument list.
111	most variables will be for integer pointers, so make integer names the default for all variables.
121-122	accommodate lack of support for **EXTERNAL** specification in MATLAB.*
127-129	put variables in **COMMON** for use by FORTRAN subprogram **odeext**.
134-141	verify the function argument lists and output an error message using MATLAB MEX-library routine **mexerrmsgtxt**. The gateway routine is aborted and control returned to calling routine if error is detected.
143-146	MATLAB MEX-library routine **mxgetm** is used to determine matrix size and length for argument **x0**.
151-154	MATLAB MEX-library routine **mxcreatefull** is used to create several matrices. Variables with the suffix _mptr, such as work_mptr, contain an integer pointer to the starting address in memory of the data structure defining the matrix.
155-158	MATLAB MEX-library routine **mxgetpr** is used to get the starting memory address for the real data elements of several matrices. These variables are identified by a suffix of _rptr, such as work_rptr. Note that the content of work_rptr is an integer pointer representing the starting address in memory where the data elements of the matrix are stored. (Complex parts are stored separately.)
	odeg *continued on next page*

	odeg *continued from previous page*
Line	Operation
159-160	prepare matrix pointers for variables to be used in FORTRAN subprogram **odeext**.
166	MATLAB MEX-library routine **mexgetstring** is used to retrieve string variable passed from M-file which invokes the gateway routine.
167-170	make sure the argument was a string.
172-178	MATLAB MEX-library routine **mexgetpr** is used to get the starting memory address for the real data elements of several matrices.
180-185	MATLAB MEX-library routine **mexgetscalar** is used to get the actual floating point value for several scalars.
186-187	change MATLAB's floating point numbers to FORTRAN integers for later use.
190-191	duplicate a MATLAB vector using FORTRAN routine **dupvector**. Note the use of the FORTRAN intrinsic function **%val** which is available with most UNIX FORTRAN compilers. **%val** is a nonstandard intrinsic used to pass variables from the gateway routine to a FORTRAN routine. Recall that the content of variable **x0_rptr** is an integer which is the starting address in memory for the data elements of this vector. Normally, FORTRAN would pass the address of variable **x0_rptr** to the subprogram. What we really wish to do is pass the value contained in **x0_ptr** to the subprogram as the starting location in memory of the data elements of the vector. **%val** accomplishes this task.
195-201	MATLAB MEX-library routine **mexcreatefull** is used to allocate storage for the output variables from the function.
	odeg *continued on next page*

odeg *continued on next page*

Line	Operation
	odeg *continued from previous page*
202-206	MATLAB MEX-library routine **mxgetpr** is used to get the starting memory address for the real data elements of several matrices.
209	duplicate a MATLAB scalar using FORTRAN routine **dupscalar**.
210-212	duplicate a column from a MATLAB matrix using FORTRAN routine **dupmatrixcol**.
214-234	perform the integration using FORTRAN routine **ode**. The user should study the documentation included with the routine **ode** to gain an understanding of the parameters.
238-243	MATLAB MEX-library routine **mxcopyreal8toptr** is used to transfer results back to MATLAB variables.
246-249	MATLAB MEX-library routine **mxfreematrix** is used to free memory allocated for temporary storage.
253-263	**ode** error condition exit.

* The authors were unable to successfully use the standard FORTRAN **EXTERNAL** specification in conjunction with MATLAB MEX-files. Presently, the MATLAB MEX-facility does not support the FORTRAN **EXTERNAL** specification. Therefore, the routine **ode** must be manually edited to use the actual function name rather than passing the function name as an argument.

TABLE 8.1. Description of FORTRAN Routine **odeg**

The user must supply a routine to evaluate the differential equation. This function is called by the integrator routine **ode**. In the example problem the M-file **equamo** serves this purpose. **equamo** is an M-file and we need to call it from a FORTRAN routine. The FORTRAN subprogram **odeext** is a FORTRAN-to-MATLAB gateway routine. This routine is called by **ode** whenever the differential equation must be evaluated. The table below describes the sequence of statements which constitute **odeext**.

odeext	
Line	Operation
22	arguments passed from routine **ode**.
28-30	variables allocated in **odeg** are transferred via COMMON.
31	specify values for number of arguments for both right and left hand side of a function call.
34-37	duplicate a vector using FORTRAN routine **dupvector**.
41-42	MATLAB MEX-library routine **mexcallmatlab** is used to invoke a MATLAB M-file from FORTRAN. Note that the parameters passed are constructed in the same fashion as those in the gateway routine **odeg**. For the example problem in this section the M-file being called is **equamo**.
45-47	save the results from this call.

TABLE 8.2. Description of FORTRAN Routine **odeext**

8.7.2.2   Substitute ODE Routine **ode.f**

The ODE routine developed by Shampine and Gordon [77] was selected as a substitute for MATLAB's **ode45** for demonstrating how to implement MEX-routines. The routine by Shampine and Gordon, **ode.f**, was written in FORTRAN and is available via the Internet. The routine can be obtained using regular Internet email as follows:

```
mail netlib@ornl.gov
  send ode from ode            >>  Body of email
  send d1mach from core        >>    message
```

The routine **d1mach** should be concatenated with **ode**. This routine is used by **ode** to determine several machine dependent constants. The user will also have to perform the following edits to routine **ode**:

- Place the letter C in column one of every line that contains an **EXTERNAL** specification (called commenting-out lines). This is required as a "work-around" to the MATLAB bug in handling **external** statements.

- Change the defined value for variable **maxnum**. This value occurs in a single **data** statement. Set the value to 2000 for this example.

- Comment-out all subprogram calls of the form "**call f(**". Replace these with an identical call statement except change the name **f** to **odeext**. This is required as a "work-around" to the MATLAB bug for handling **EXTERNAL** statements.

- In subprogram **d1mach** un-comment-out the set of statements which match your hardware system.

8.7.2.3 Compiling MEX Routines

The standard MATLAB distribution for UNIX systems includes a script (batch file) called **fmex** which creates the necessary MEX object file. The statement below will compile, link, and create the required MEX object file for this example. The **-O** option instructs the compiler to perform optimization on the FORTRAN code.

```
fmex -O ode.f odeg.f odeext.f dup.f
```

8.7.3 DISCUSSION OF RESULTS FROM USING A MEX-ROUTINE

An example involving differential equation solvers was selected to illustrate implementation of MEX-files. An Adams type integrator was used, since this method is a popular alternative to Runge-Kutta methods. Of course, the appeal of any integrator depends on its accuracy and speed. We chose **ode.f** (.f and .m extensions are adopted to distinguish between FORTRAN and MATLAB) because it is readily available, has a relatively small number of source lines, and is fully described by Shampine and Gordon [77]. Other more versatile and comprehensive codes such as *DASL* [9], for example, are several thousand lines long.

Execution speed in the cable dynamics problem depends on whether the code is computed or interpreted, the number of code lines being executed, and the quality of the algorithm used. Generally, compiled code runs from five to twenty times faster than interpreted code. However, the number of source lines in **ode45.m** is small, so not having this function compiled may not affect performance greatly. Nearly all of the computation in this example occurs in function **equamo.m** which utilizes various matrix manipulative features such as matrix addition, matrix multiplication, **flipud**, **cumsum**, **tril**, and the \ operator for simultaneous equation solution. Matrix multiplication and equation solution are compiled functions in MATLAB and, therefore, should execute as quickly as compiled FORTRAN. Of course, converting the MATLAB code in **equamo.m** to FORTRAN would also produce substantial speed improvement. This assertion has been verified by our previous experience in solving cable problems completely in FORTRAN. Evidently, MATLAB has a great appeal because it leads to solutions which are concise and can be implemented quickly using substantially fewer lines of code than is necessary with a language like FORTRAN.

Let us compare how well **ode45.m** and **ode.f** solved the cable dynamics problem. Both codes required similar computation times to achieve comparable accuracy, but we found **ode45.m** to be more accurate than **ode.f**, for this example. When using **ode.f**, we tried several combinations of integration tolerance and different choices of intervals at which output was produced. The best results obtained from **ode.f** were consistently less accurate than output from **ode45.m**. Results from **ode.f** for several choices of integration stepsize and number of output intervals appear in Table 8.7.3. Figures 8.12 and 8.13 present **ode.f** results for midpoint deflection and error growth (third parameter set from Table 8.7.3) which are similar to Figures 8.9 and 8.10 produced using **ode45.m**. Although **ode.f** required slightly less computation time, solution accuracy was lower. The Adams integrator results went unstable by $t = 15$, whereas the Runge-Kutta results were still accurate up to $t = 20$. Furthermore, tightening the integration tolerance in **ode.f** greatly increased computation time without producing much improvement in solution accuracy.

relerr	abserr	ngrid	CPU time (secs)
1E-6	1E-6	100	161.55
1E-6	1E-6	500	161.46
1E-6	1E-6	1000	161.74
1E-9	1E-9	1000	277.16
Note: M-file only implementation required 176.08 seconds of CPU time.			

TABLE 8.3. M-file and MEX Timing Comparison

We end this chapter with a caveat that results from our cable dynamics example do not imply that Runge-Kutta integrators are always superior to Adams integrators. Undoubtedly, the quality of results obtained will depend on the type of problem being solved and the choices the user makes of various code parameters.

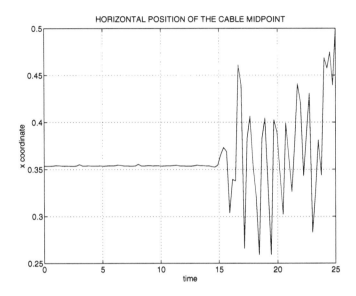

FIGURE 8.12. Horizontal Position of the Cable Midpoint

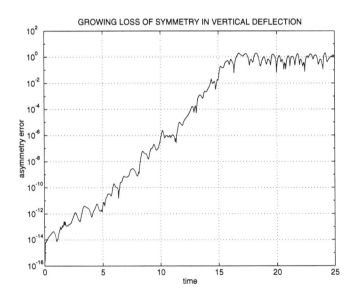

FIGURE 8.13. Growing Loss of Symmetry in Vertical Deflection

8.7.3.1   MEX Gateway Routine odeg.f

```
 1: *-------------------------------------------------
 2: *FUNCTION: Fortran MEX gateway routine
 3: *.........................................
 4: *MATLAB CALL: [t,x,rerr,aerr,istat] = ode(
 5: *                m_name, t0, tf, x0, relerr,
 6: *                abserr, ngrid, iflag )
 7: *.........................................
 8: *INPUT: iflag,   =+1, normal
 9: *                =-1, set negative only if it is
10: *                     impossible to continue the
11: *                     integration beyond tf
12: *       ngrid    number of increments between t0
13: *                and tf to use
14: *       m_name   name of m-file defining the
15: *                DEs to be integrated
16: *       t0       starting time for integration
17: *       tf       ending time for integration
18: *       x0       vector of initial conditions
19: *       relerr   relative error tolerance for
20: *                local error test
21: *       abserr   absolute error tolerance for
22: *                local error test,
23: *
24: *                [ abs(local err) .le.
25: *                    abs(x)*relerr + abserr ]
26: *.........................................
27: *OUTPUT: t       last time reached in
28: *                integration (normally = tf)
29: *        x       solution at t
30: *        rerr    new relative error tolerance
31: *                when istat = 3
32: *        aerr    new absolute error tolerance
33: *                when istat = 3
34: *        istat   =2, normal, integration
35: *                    reached tf
36: *                =3, did not reach tf, error
37: *                    tolerances too small,
38: *                    (relerr, abserr) increased
39: *                    appropriately
40: *                =4, did not reach tf, >500
41: *                    steps needed
42: *                =5, did not reach tf, equations
```

```
43: *                    appear to be too stiff
44: *                    =6, invalid input parameters
45: *                    =-, values are negative if
46: *                    iflag was set negative
47: *.........................................
48: *STRUCTURE: Given the MATLAB construct for
49: *           a function as:
50: *               [a,b,c,...] = fun(d,e,f,...)
51: *
52: *           NLHS, number of quantities on
53: *                 left side
54: *                 (e.g. for [a,b], NLHS = 2)
55: *           plhs, address pointer vector
56: *                 NLHS pointers in vector plhs
57: *           NRHS, number of quantities on
58: *                 right side
59: *                 (e.g. for [d,e,f], NRHS = 3)
60: *           prhs, address pointer vector
61: *                 NRHS pointers in vector prhs
62: *.........................................
63: *CALLS: Intrinsic: %val, dfloat, int
64: *       User:      ode, dupmatrixcol,
65: *                  dupscalar, dupvector
66: *       Matlab:    mexerrmsgtxt,
67: *                  mxcopyreal8toptr,
68: *                  mxcreatefull, mxfreematrix,
69: *                  mxgetm, mxgetn, mxgetpr,
70: *                  mxgetscalar, mxgetstring
71: *.........................................
72: *NOTES: * This is a MATLAB Mex gateway
73: *         routine for the routine ode.f which
74: *         is available from netlib.  More
75: *         information about the netlib routine
76: *         can be found in the text "Computer
77: *         Solution of Ordinary Differential
78: *         Equations" by L.F. Shampine and
79: *         M.K. Gordon.  To retrieve these
80: *         routines send the following internet
81: *         email message:
82: *
83: *         mail netlib@ornl.gov
84: *             send ode from ode
85: *             send d1mach from core
86: *
87: *         You must edit these files to setup
```

```
 88: *              some machine specific values, etc.
 89: *
 90: *          * Use of %val intrinsic.  This
 91: *            intrinsic function is an extension
 92: *            to standard Fortran-77.  It is NOT
 93: *            supported on all platforms.  A
 94: *            MATLAB related discussion of %val
 95: *            is contained in the MATLAB External
 96: *            Interface Guide in the section
 97: *            titled:
 98: *              "Details About Fortran MEX-Files"
 99: *
100: *          * Program flow:
101: *
102: *      MATLAB (main calling routine)
103: *      |> odeg.f (mexfunction gateway routine)
104: *        |> ode.f (netlib ODE routine)
105: *          |> odeext.f (user external function)
106: *            |> XXXX.m (user matlab function)
107: *
108: *-------------------------------------------------
109:        subroutine mexfunction( nlhs, plhs,
110:      &                          nrhs, prhs )
111:        implicit    integer (a-z)
112:        integer     e_prhs(2), plhs(*), prhs(*),
113:      &             i_work(5)
114:        real*8      mxgetscalar
115:        real*8      r_abserr, r_iflag, r_ngrid,
116:      &             r_relerr, r_t0, r_t0_loop,
117:      &             r_tf, r_tf_loop, r_tinc,
118:      &             r_zero
119:        character   m_name*24
120: C-*
121:        character   ext_name*8
122: C      external    ext_name
123: C-*
124: C...Used to pass critical pointers to gateway
125: C...routine back into MATLAB m-function
126: C...describing DEs
127:        common      / share / m_name, nlen,
128:      &                       e_prhs, ytmp_rptr,
129:      &                       ttmp_rptr
130: C...Name of external routine called from ode
131:        data        ext_name / 'odeext' /
132:        data        r_zero/ 0.0d0 /
```

```
133:
134: C...Verify MATLAB calling routine argument list
135:      if( nrhs .ne. 8 ) then
136:         call mexerrmsgtxt
137:    &  ( 'mex ODE error: 8 input arguments?' )
138:      elseif( nlhs .ne. 5 ) then
139:         call mexerrmsgtxt
140:    &  ('mex ODE error: 5 output arguments?' )
141:      endif
142:
143: C...Get matrix size for x0
144:      m    = mxgetm( prhs(4) )
145:      n    = mxgetn( prhs(4) )
146:      nlen = m * n
147:
148: C...Create temporary work arrays and assign
149: C...pointers
150:      m           = 100 + 21 * nlen
151:      work_mptr = mxcreatefull( m, 1, 0 )
152:      xtmp_mptr = mxcreatefull( nlen, 1, 0 )
153:      ytmp_mptr = mxcreatefull( nlen, 1, 0 )
154:      ttmp_mptr = mxcreatefull( 1, 1, 0 )
155:      work_rptr = mxgetpr( work_mptr )
156:      xtmp_rptr = mxgetpr( xtmp_mptr )
157:      ytmp_rptr = mxgetpr( ytmp_mptr )
158:      ttmp_rptr = mxgetpr( ttmp_mptr )
159:      e_prhs(1) = ttmp_mptr
160:      e_prhs(2) = ytmp_mptr
161:
162: C...Get input parameters
163:      m     = mxgetm( prhs(1) )
164:      n     = mxgetn( prhs(1) )
165:      m_name = ' '
166:      ichk = mxgetstring( prhs(1), m_name, m*n)
167:      if( ichk .eq. 1 ) then
168:         call mexerrmsgtxt
169:    &  ('mex ODE error: m_name not a string')
170:      endif
171:
172:      t0_rptr     = mxgetpr( prhs(2) )
173:      tf_rptr     = mxgetpr( prhs(3) )
174:      x0_rptr     = mxgetpr( prhs(4) )
175:      relerr_rptr = mxgetpr( prhs(5) )
176:      abserr_rptr = mxgetpr( prhs(6) )
177:      ngrid_rptr  = mxgetpr( prhs(7) )
```

```
178:          iflag_rptr  = mxgetpr( prhs(8) )
179:
180:          r_t0      = mxgetscalar( prhs(2) )
181:          r_tf      = mxgetscalar( prhs(3) )
182:          r_relerr  = mxgetscalar( prhs(5) )
183:          r_abserr  = mxgetscalar( prhs(6) )
184:          r_ngrid   = mxgetscalar( prhs(7) )
185:          r_iflag   = mxgetscalar( prhs(8) )
186:          i_ngrid   = int(r_ngrid)
187:          i_iflag   = int(r_iflag)
188:
189: C... transfer parameters
190:          call dupvector( %val(x0_rptr),
191:     &                    %val(xtmp_rptr), nlen )
192:
193: C...Create required output arrays and
194: C...assign pointers
195:          plhs(1)   = mxcreatefull
196:     &              ( i_ngrid+1, 1, 0 )
197:          plhs(2)   = mxcreatefull
198:     &              ( i_ngrid+1, nlen, 0 )
199:          plhs(3)   = mxcreatefull( 1, 1, 0 )
200:          plhs(4)   = mxcreatefull( 1, 1, 0 )
201:          plhs(5)   = mxcreatefull( 1, 1, 0 )
202:          t_rptr    = mxgetpr( plhs(1) )
203:          x_rptr    = mxgetpr( plhs(2) )
204:          rerr_rptr = mxgetpr( plhs(3) )
205:          aerr_rptr = mxgetpr( plhs(4) )
206:          istat_rptr = mxgetpr( plhs(5) )
207:
208: C...Make boundary conditions solution at t=t0
209:          call dupscalar( r_zero, %val(t_rptr), 1 )
210:          call dupmatrixcol( %val(x0_rptr),
211:     &                       %val(x_rptr),
212:     &                       1, nlen, i_ngrid+1 )
213:
214: C...Integrate over time
215:          r_tinc    = ( r_tf - r_t0 ) / r_ngrid
216:          r_t0_loop = r_t0
217:          do 100 i = 2, i_ngrid+1
218:            r_tf_loop = ( i - 1 ) * r_tinc + r_t0
219:            call ode( ext_name, nlen,
220:     &                %val(xtmp_rptr), r_t0_loop,
221:     &                r_tf_loop, r_relerr,
222:     &                r_abserr, i_iflag,
```

```
223:    &                    %val(work_rptr), i_work )
224:          if( abs(i_iflag) .ge. 3 .and.
225:    &          abs(i_iflag) .le. 6 ) go to 999
226:
227: C...Transfer results back
228:          call dupscalar( r_t0_loop,
229:    &                    %val(t_rptr), i )
230:          call dupmatrixcol( %val(xtmp_rptr),
231:    &                       %val(x_rptr),
232:    &                       i, nlen, i_ngrid+1 )
233:
234:  100     continue
235:
236: C...Wrap up
237:        r_iflag = dfloat(i_iflag)
238:        call mxcopyreal8toptr
239:    &         ( r_relerr, rerr_rptr, 1 )
240:        call mxcopyreal8toptr
241:    &         ( r_abserr, aerr_rptr, 1 )
242:        call mxcopyreal8toptr
243:    &         ( r_iflag, istat_rptr, 1 )
244:
245: C...Free temporary storage
246:        call mxfreematrix( work_mptr )
247:        call mxfreematrix( xtmp_mptr )
248:        call mxfreematrix( ytmp_mptr )
249:        call mxfreematrix( ttmp_mptr )
250:
251:        return
252:
253: C...Error condition exit
254:  999     continue
255:        r_iflag = dfloat(i_iflag)
256:        call mxcopyreal8toptr
257:    &         ( r_relerr, rerr_rptr, 1 )
258:        call mxcopyreal8toptr
259:    &         ( r_abserr, aerr_rptr, 1 )
260:        call mxcopyreal8toptr
261:    &         ( r_iflag, istat_rptr, 1 )
262:        print *, 'istat = ', i_iflag
263:        call mexerrmsgtxt('mex ODE error: istat' )
264:
265:        return
266:        end
```

8.7.3.2   MEX Routine odeext.f

```
 1: *-------------------------------------------------
 2: *FUNCTION: User defined external routine for
 3: *          ode. This routine serves as a
 4: *          gateway routine to call back to a
 5: *          MATLAB M-file which performs
 6: *          dy(i)/dt.
 7: *.............................................
 8: *FORTRAN CALL: CALL ODEEXT( T, Y, YP )
 9: *.............................................
10: *INPUT: T, time
11: *       Y, vector y(i)
12: *.............................................
13: *OUTPUT: YP, vector y'(i)
14: *.............................................
15: *CALLS: Intrinsic: %val
16: *        User:      dupvector
17: *        Matlab:    mexcallmatlab, mxgetpr,
18: *                   m-routine with name m_name
19: *.............................................
20: *NOTES: See odeg for discussion of %val use.
21: *-------------------------------------------------
22:        subroutine odeext( t, y, yp )
23:        implicit   real*8(a-h,o-z)
24:        real*8     y(*), yp(*)
25:        integer    e_plhs(1), e_prhs(2),
26:       &           ttmp_rptr, yp_rptr, ytmp_rptr
27:        character  m_name*24
28:        common     / share / m_name, nlen,
29:       &                     e_prhs, ytmp_rptr,
30:       &                     ttmp_rptr
31:        data       mrhs/ 2 /, mlhs/ 1 /
32:
33: C...Transfer to temporary MATLAB variables
34:        call dupvector
35:       & ( y,  %val(ytmp_rptr),  nlen )
36:        call dupvector
37:       & ( t,  %val(ttmp_rptr),  1 )
38:
39: C...Call a MATLAB M-file function with form:
40: C... [yp]=m_name(t,y)
41:        call mexcallmatlab
42:       & ( mlhs, e_plhs, mrhs, e_prhs, m_name)
```

```
43:
44: C...Transfer/save the results
45:       yp_rptr  = mxgetpr( e_plhs(1) )
46:       call dupvector
47:     & ( %val(yp_rptr), yp, nlen )
48:
49:       return
50:       end
```

8.7.3.3   MEX Routine dup.f

```
 1: *---------------------------------------------------
 2: *FUNCTION: Duplicate MATLAB tensors
 3: *...............................................*..
 4: *FORTRAN CALL: CALL DUPSCALAR(    A, B, KPOS )
 5: *              CALL DUPVECTOR(    A, B, NLEN )
 6: *              CALL DUPMATRIXCOL( A, B, KROW,
 7: *                                 NLEN, MXROW )
 8: *..............................................*..
 9: *INPUT: A,     source tensor for duplication
10: *       KPOS,  index in B to place A
11: *       KROW,  row index in B to place A
12: *       NLEN,  length of vector tensor to
13: *              duplication
14: *       MXROW, row dimension of matrix for
15: *              dimensioning purposes
16: *...............................................*.
17: *OUTPUT: B, destination tensor of duplication
18: *...............................................*.
19: *CALLS: Intrinsic: None.
20: *       User:      None.
21: *       Matlab:    None.
22: *..............................................*..
23: *NOTES: These routines are used to copy tensors
24: *       which are defined by MATLAB pointers.
25: *---------------------------------------------------
26:       subroutine dupscalar( a, b, kpos )
27:       real*8    a, b(*)
28:       b(kpos) = a
29:       return
30:       end
31: *---------------------------------------------------
32:       subroutine dupvector( a, b, nlen )
```

```
33:        real*8      a(*), b(*)
34:        do 10 i = 1, nlen
35:   10     b(i) = a(i)
36:        return
37:        end
38: *-----------------------------------------------
39:        subroutine dupmatrixcol( a, b, krow,
40:       &                          nlen, mxrow )
41:        real*8      a(nlen), b(mxrow,nlen)
42:        do 10 i = 1, nlen
43:   10     b(krow,i) = a(i)
44:        return
45:        end
```

9

Boundary Value Problems for Linear Partial Differential Equations

9.1 Several Important Partial Differential Equations

Many physical phenomena are characterized by linear partial differential equations. Such equations are attractive to study because (a) principles of superposition apply in the sense that linear combinations of component solutions can often be used to build more general solutions (b) finite difference or finite element approximations lead to systems of linear equations amenable to solution by matrix methods. The table below lists several frequently encountered equations and some applications. We only show one- or two-dimensional forms, although some of these equations have relevant applications in three dimensions.

In most practical applications the differential equations must be solved within a finite region of space while simultaneously prescribing boundary conditions on the function and its derivatives. Furthermore, initial conditions may exist. In dealing with the initial value problem, we are trying to predict future system behavior when initial conditions, boundary conditions, and a governing physical process are known. Solutions to such problems are seldom

Equation	Equation Name	Applications
$u_{xx} + u_{yy} = \alpha u_t$	Heat	Transient heat conduction
$u_{xx} + u_{yy} = \alpha u_{tt}$	Wave	Transverse vibrations of membranes and other wave type phenomena
$u_{xx} + u_{yy} = 0$	Laplace	Steady-state heat conduction and electrostatics
$u_{xx} + u_{yy} = f(x, y)$	Poisson	Stress analysis of linearly elastic bodies
$u_{xx} + u_{yy} + \omega^2 u = 0$	Helmholtz	Steady-state harmonic vibration problems
$EI y_{xxxx} = -A\zeta y_{tt} + f(x, t)$	Beam	Transverse flexural vibrations of elastic beams

obtainable in a closed finite form. Even when series solutions are developed, an infinite number of terms may be needed to provide generality. For example, the problem of transient heat conduction in a circular cylinder leads to an infinite series of Bessel functions employing characteristic values which can only be computed approximately. Hence, the notion of an "exact" solution expressed as an infinite series of transcendental functions is deceiving. At best, we can hope to produce results containing insignificantly small computation errors.

The present chapter studies five problems. Four of the five problems are solved by series methods. The last problem is distinctive from the others since exact natural frequencies of an elastic beam are compared with approximations produced by finite difference and finite element methods.

9.2 Solving the Laplace Equation Inside a Rectangular Region

Functions which satisfy Laplace's equation are encountered often in practice. Such functions are called harmonic; and the problem of determining a harmonic function subject to given boundary values is known as the Dirichlet problem [99]. In a few cases with simple geometries, the Dirichlet problem can be solved explicitly. One instance is a rectangular region with the boundary values of the function being expandable in a Fourier sine series. The following program employs the FFT to construct a solution for boundary values represented by piecewise linear interpolation. Surface and contour plots of the resulting field values are also presented.

The problem of interest satisfies the differential equation

$$\frac{\partial^2 u}{\partial x^2} + \frac{\partial^2 u}{\partial y^2} = 0 \qquad 0 < x < a \qquad 0 < y < b$$

with the boundary conditions of the form

$$
\begin{aligned}
u(x,0) &= F(x) & 0 < x < a \\
u(x,b) &= G(x) & 0 < x < a \\
u(0,y) &= P(y) & 0 < y < b \\
u(a,y) &= Q(y) & 0 < y < b
\end{aligned}
$$

The series solution can be represented as

$$u(x,y) = \sum_{n=1}^{\infty} f_n a_n(x,y) + g_n a_n(x, b-y) +$$

$$p_n b_n(x,y) + q_n b_n(a-x,y)$$

where

$$a_n(x,y) = \sin\left[\frac{n\pi x}{a}\right] \sinh\left[\frac{n\pi(b-y)}{a}\right] / \sinh\left[\frac{n\pi b}{a}\right]$$

$$b_n(x,y) = \sinh\left[\frac{n\pi(a-x)}{b}\right] \sin\left[\frac{n\pi y}{b}\right] / \sinh\left[\frac{n\pi a}{b}\right]$$

and the constants f_m, g_m, p_n, and q_n are coefficients in the Fourier sine expansions of the boundary value functions. This implies that

$$F(x) = \sum_{n=1}^{\infty} f_n \sin\left[\frac{n\pi x}{a}\right] \qquad G(x) = \sum_{n=1}^{\infty} g_n \sin\left[\frac{n\pi x}{a}\right]$$

$$P(y) = \sum_{n=1}^{\infty} p_n \sin\left[\frac{n\pi y}{b}\right] \qquad Q(y) = \sum_{n=1}^{\infty} q_n \sin\left[\frac{n\pi y}{b}\right]$$

The coefficients in the series can be computed by integration as

$$f_n = \frac{2}{a} \int_0^a F(x) \sin\left[\frac{n\pi x}{a}\right] dx \qquad g_n = \frac{2}{a} \int_0^a G(x) \sin\left[\frac{n\pi x}{a}\right] dx$$

$$p_n = \frac{2}{a} \int_0^b P(y) \sin\left[\frac{n\pi y}{b}\right] dy \qquad q_n = \frac{2}{a} \int_0^b Q(y) \sin\left[\frac{n\pi y}{b}\right] dy$$

or approximate coefficients can be obtained using the FFT. The latter approach is chosen here and the solution is evaluated for an arbitrary number of terms in the series.

The chosen problem solution has the disadvantage of employing eigenfunctions which vanish at the ends of the expansion intervals. Consequently, it is desirable to combine the series with an additional term allowing exact satisfaction of the corner conditions for cases where the boundary value functions for adjacent corners agree. This implies requirements such as $F(a) = Q(0)$ and three other similar conditions. It is evident that the function

$$u_p(x,y) = c_1 + c_2 x + c_3 y + c_4 xy$$

is harmonic and varies linearly along each side of the rectangle. Constants $c_1, \cdots, c_4$ can be computed to satisfy the corner values and the total solution is represented as u_p plus a series solution involving modified boundary conditions.

The following program solves the boundary value problem and plots results. The functions used in this program are described below.

laplarec	sets data parameters, establishes global variables, calls other functions, and generates output
setup	generates coefficients defining a particular solution to satisfy corner values
ulbc	the harmonic function pertaining to corner conditions
sinfft	generates coefficients in a Fourier sine series
laprec	sums the series solution of Laplace's equation
lintrp	piecewise linear interpolation function
fbot,gtop,plft,qrht	boundary value functions for bottom, top, left side, and right side of the rectangle. These are specified as piecewise linear.
read	utility function to read data
genprint	utility function to save graph to file
removfil	utility function to delete a file

The example data set defined in the driver program has no particular significance other than to produce interesting surface and contour plots. Different boundary conditions can be handled by slight modifications of the input data.

In this example only 30 series terms were used to illustrate how Fourier series based solutions behave near discontinuities in boundary values. Oscillations in solution values occur at corners and other points where jump discontinuities occur. This behavior is typical of Fourier series discussed in Chapter 6.

SURFACE PLOT

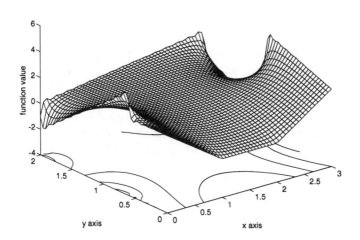

FIGURE 9.1. Surface Plot

CONTOUR PLOT

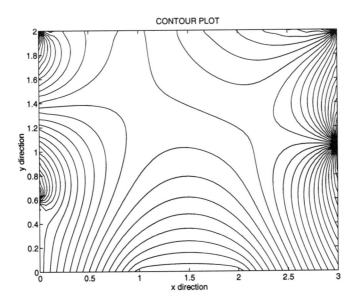

FIGURE 9.2. Contour Plot

MATLAB EXAMPLE

9.2.0.1 Script file laplarec

```
 1: % Example:  laplarec
 2: % ~~~~~~~~~~~~~~~~~~
 3: % This program uses Fourier series methods
 4: % to solve Laplace's equation in a rectangle.
 5: % Boundary conditions are defined by piecewise
 6: % linear interpolation of boundary data. The
 7: % program can easily be changed to deal with
 8: % problems where the boundary conditions are
 9: % expressed by analytically defined functions.
10: % Surface and contour plots of the solution
11: % are also provided.
12: %
13: % User m functions required:
14: %    setup, ulbc, sinfft, laprec, fbot, gtop,
15: %    plft, qrht, lintrp, genprint, removfil
16: %-------------------------------------------------
17:
18: clear
19: global a_ b_ xbot_ ubot_ xtop_ utop_
20: global ylft_ ulft_ yrht_ urht_
21: global u1_ u2_
22:
23: a_=3; b_=2; % Rectangle side lengths
24:
25: % Data for piecewise linear
26: % boundary conditions
27: xtop_=[0;    3];    utop_= [0;    2];
28: ntop=length(xtop_);
29: xbot_=[0;1;2;3];    ubot_=[2;-1;-1;2];
30: nbot=length(xbot_);
31: ylft_=[0;0.6;0.6;2]; ulft_=[2;2;4;-2];
32: nlft=length(ylft_);
33: yrht_=[0;1;1.1;2];  urht_=[2;4;-2; 0];
34: nrht=length(ylft_);
35:
36: % Create the constants needed in
37: % function ulbc
38: cc=setup;
39:
40: % Adjust boundary data to give zero
```

```
41: % end conditions
42: for k=1:nbot
43:    ubot_(k)=ubot_(k)-ulbc(cc,xbot_(k),0);end
44: for k=1:ntop
45:    utop_(k)=utop_(k)-ulbc(cc,xtop_(k),b_);end
46: for k=1:nlft
47:    ulft_(k)=ulft_(k)-ulbc(cc,0,ylft_(k));end
48: for k=1:nrht
49:    urht_(k)=urht_(k)-ulbc(cc,a_,yrht_(k));end
50:
51: % Generate Fourier coefficients for the
52: % modified boundary conditions
53: sbot=sinfft('fbot',a_,9);
54: stop=sinfft('gtop',a_,9);
55: slft=sinfft('plft',b_,9);
56: srht=sinfft('qrht',b_,9);
57:
58: % Generate a grid of interior points
59: % for solution evaluation
60: nin=51; ntrms=30;
61: xin=linspace(0,a_,nin); yin=linspace(0,b_,nin);
62:
63: % Evaluate the solution having zero
64: % corner values
65: uin=laprec(...
66:     sbot,stop,slft,srht,a_,b_,xin,yin,ntrms);
67:
68: % Correct results for nonzero corner values
69:
70: uin=uin+ulbc(cc,xin,yin); uin=flipud(uin');
71:
72: % Display surfaces showing function
73: % values on the grid
74:
75: meshc(xin,yin,flipud(uin));
76: xlabel('x axis'); ylabel('y axis');
77: zlabel('function value');
78: title('SURFACE PLOT')
79: disp('Press RETURN to continue'); pause;
80: genprint('lapsrfac'); clf
81:
82: contour(xin,yin,flipud(uin),30)
83: title('CONTOUR PLOT')
84: xlabel('x direction'); ylabel('y direction');
85: genprint('contur');
```

```
86: disp('All Done')
```

9.2.0.2   Function setup

```
 1: function cc=setup
 2: %
 3: % cc=setup
 4: % ~~~~~~~~
 5: % cc - coefficients used to define the
 6: %       particular solution to satisfy
 7: %       corner conditions
 8: %
 9: % User m functions called:
10: %    fbot, qrht, plft, gtop
11: %------------------------------------------------
12:
13: global  a_ b_
14:
15: s=(a_+b_)/1e10;
16: c1=(fbot(s)+plft(s))/2;
17: c2=(fbot(a_-s)+qrht(s))/2;
18: c3=(plft(b_-s)+gtop(s))/2;
19: c4=(gtop(a_-s)+qrht(b_-s))/2;
20: mat=[[1,0,0,0];[1,a_,0,0];
21:      [1,0,b_,0];[1,a_,b_,a_*b_]];
22: vec=[c1;c2;c3;c4];  cc=mat\vec;
```

9.2.0.3   Function ulbc

```
 1: function u=ulbc(c,x,y)
 2: %
 3: % u=ulbc(c,x,y)
 4: % ~~~~~~~~~~~~~
 5: % This function determines a harmonic function
 6: % satisfying boundary conditions which vary
 7: % linearly on the sides of a rectangle.
 8: %
 9: % User m functions called:  lintrp
10: %-----------------------------------------------
11:
12: x=x(:); y=y(:)'; nx=length(x); ny=length(y);
```

```
13: x=x*ones(1,ny); y=ones(nx,1)*y;
14: u=c(1)*ones(nx,ny)+c(2)*x+c(3)*y+c(4)*x.*y;
```

9.2.0.4   Function sinfft

```
 1: function sincof = sinfft(fun,hafper,powr2)
 2: %
 3: % sincof = sinfft(fun,hafper,powr2)
 4: % ~~~~~~~~~~~~~~~~~~~~~~~~~~~~~~~~~~~~
 5: % This function determines coefficients in
 6: % the Fourier sine series of a general real
 7: % valued function.
 8: %
 9: %   hafper - the half  period over which the
10: %            expansion applies
11: %   fun    - the real valued function being
12: %            expanded. This function must be
13: %            defined for vector arguments with
14: %            components between zero
15: %            and hafper.
16: %   powr2  - the power of 2 determining the
17: %            number of function values used
18: %            in the FFT (number of
19: %            values = 2^powr2).  When powr2
20: %            is 9, then 512 function values
21: %            are used and 255 Fourier
22: %            coefficients are computed.
23: %
24: % User m functions called:  none
25: %-----------------------------------------------
26:
27: n=2^powr2; period=2*hafper; n2=n/2;
28: x=(period/n)*(0:n2);
29: fval=feval(fun,x);
30: fval=fval(:); fval=[fval;-fval(n2:-1:2)];
31: foucof=fft(fval);
32: sincof=-(2/n)*imag(foucof(2:n2));
```

9.2.0.5   Function laprec

```
 1: function u=laprec(f,g,p,q,a,b,x,y,ntrms)
```

```
2: %
3: % u=laprec(f,g,p,q,x,y,ntrms)
4: % ~~~~~~~~~~~~~~~~~~~~~~~~~~~~
5: % This function sums the series which solves
6: % Laplace's equation in a rectangle.
7: %
8: %   f,g,p,q - Fourier sine series coefficients
9: %             for the boundary conditions on
10: %            the bottom, top, left, and
11: %            right sides
12: %   a,b     - the horizontal and vertical
13: %            side lengths
14: %   x,y     - vectors containing coordinates
15: %            of points defining a rectangular
16: %            grid on which the solution is
17: %            to be evaluated.
18: %   ntrms   - number of series terms used (not
19: %            exceeding length(f));
20: %
21: % User m functions called:  none
22: %--------------------------------------------
23:
24: nt=length(f);
25: if nargin==8, ntrms=nt; end;
26: ntrms=min(nt,ntrms);
27: x=x(:); y=y(:)';
28: n=1:ntrms; nx=length(x); ny=length(y);
29: if nt>ntrms
30:   f=f(n); g=g(n); p=p(n); q=q(n); end
31: a2=2*a; b2=2*b; na=(pi/a)*n; nb=(pi/b)*n;
32: denomx=1-exp(-b2*na(:));
33: f=f(:)./denomx; g=g(:)./denomx;
34: denomy=1-exp(-a2*nb(:)');
35: p=p(:)'./denomy; q=q(:)'./denomy;
36:
37: u1_=(f*ones(1,ny)).* ...
38:     (exp(-na'*y)-exp(-na'*(b2-y)));
39: u2_=(g*ones(1,ny)).* ...
40:     (exp(-na'*(b-y))-exp(-na'*(b+y)));
41: u=sin(x*na)*( u1_+u2_ );
42: u3_=(exp(-x*nb)-exp(-(a2-x)*nb)).* ...
43:     (ones(nx,1)*p);
44: u4_=(exp(-(a-x)*nb)-exp(-(a+x)*nb)).* ...
45:     (ones(nx,1)*q);
46: u=u+(u3_+u4_)*sin(nb'*y);
```

9.2.0.6   Function fbot

```
1: function ubot=fbot(x)
2: %
3: % ubot=fbot(x)
4: % ~~~~~~~~~~~~~
5: % x    - vector argument
6: % ubot - function value on bottom side
7: %
8: % User m functions called:  lintrp
9: %------------------------------------------------
10:
11: global xbot_ ubot_
12:
13: ubot=lintrp(x,xbot_,ubot_);
```

9.2.0.7   Function gtop

```
1: function utop=gtop(x)
2: %
3: % utop=gbot(x)
4: % ~~~~~~~~~~~~~
5: % x    - vector argument
6: % gtop - function value on top side
7: %
8: % User m functions called:  lintrp
9: %------------------------------------------------
10:
11: global xtop_ utop_
12:
13: utop=lintrp(x,xtop_,utop_);
```

9.2.0.8   Function plft

```
1: function ulft=plft(y)
2: %
3: % ulft=plft(y)
4: % ~~~~~~~~~~~~~
5: % y    - vector argument
6: % ulft - function value on left side
```

```
 7: %
 8: % User m functions called:  lintrp
 9: %---------------------------------------------
10:
11: global ylft_ ulft_
12:
13: ulft=lintrp(y,ylft_,ulft_);
```

9.2.0.9 Function qrht

```
 1: function urht=qrht(y)
 2: %
 3: % urht=qrht(y)
 4: % ~~~~~~~~~~~~
 5: % y    - vector argument
 6: % urht - function value on right side
 7: %
 8: % User m functions called:  lintrp
 9: %---------------------------------------------
10:
11: global yrht_ urht_
12:
13: urht=lintrp(y,yrht_,urht_);
```

9.3   Transient Heat Conduction in a One-Dimensional Slab

Let us analyze the temperature history in a slab which has the left side insulated while the right side temperature varies sinusoidally according to $U_0 \sin(\Omega T)$. The initial temperature in the slab is specified to be zero. The pertinent boundary value problem is

$$\alpha \frac{\partial^2 U}{\partial X^2}(X, T) = \frac{\partial U}{\partial T}(X, T) \qquad 0 < X < \ell \qquad T > 0$$

$$\frac{\partial U}{\partial X}(0, T) = 0 \qquad U(\ell, T) = U_0 \sin(\Omega T)$$

$$U(X, 0) = 0 \qquad 0 < X < \ell$$

where U, X, T, and α are, respectively, temperature, position, time, and thermal diffusivity.

The problem can be converted to dimensionless form by taking

$$u = U/U_0 \qquad x = X/\ell \qquad t = \alpha T/\ell^2 \qquad \omega = \Omega \ell^2/\alpha$$

Then we get

$$\frac{\partial^2 u}{\partial x^2} = \frac{\partial u}{\partial x} \qquad 0 < x < 1 \qquad t > 0$$

$$\frac{\partial u}{\partial x}(0, t) = 0 \qquad u(1, t) = \mathbf{imag}\left(e^{i\omega t}\right) \qquad u(x, 0) = 0$$

The solution consists of two parts as $u = w + v$ where w is a particular solution satisfying the differential equation and non-homogeneous boundary conditions, and v is a solution satisfying homogeneous boundary conditions and specified to impose the desired zero initial temperature when combined with w. The appropriate form for the particular solution is

$$w = \mathbf{imag}\left[f(x)e^{i\omega t}\right]$$

Making w satisfy the heat equation requires

$$f''(x) = \imath \omega f(x).$$

Consequently

$$f(x) = c_1 \sin(\phi x) + c_2 \cos(\phi x)$$

where $\phi = \sqrt{-\imath\omega}$. The conditions of zero gradient at $x = 0$ and unit function value at $x = 1$ determine c_1 and c_2. We get the particular solution as

$$w = \mathbf{imag}\left[\frac{\cos(\phi x)}{\cos(\phi)}e^{\imath\omega t}\right]$$

This forced response solution evaluated at $t = 0$ yields

$$w(x, 0) = \mathbf{imag}\left[\frac{\cos(\phi x)}{\cos(\phi)}\right]$$

The general solution of the heat equation satisfying zero gradient at x=0 and zero function value at $x = 1$ is found to be

$$v(x, t) = \sum_{n=1}^{\infty} a_n \cos(\lambda_n x)e^{-\lambda_n^2 t}$$

when $\lambda_n = \pi(2n - 1)/2$. To make the initial temperature equal zero in the combined solution, the coefficients a_n are chosen to satisfy

$$\sum_{n=1}^{\infty} a_n \cos(\lambda_n x) = -\mathbf{imag}\left[\frac{\cos(\phi x)}{\cos(\phi)}\right]$$

The orthogonality of the functions $\cos(\lambda_n x)$ implies

$$a_n = -2\int_0^1 \mathbf{imag}\left[\frac{\cos(\phi x)}{\cos(\phi)}\right]\cos(\lambda_n x)dx$$

which can be integrated to give

$$a_n = -\mathbf{imag}\left[\frac{(\sin(\lambda_n + \phi)/(\lambda_n + \phi) + \sin(\lambda_n - \phi)/(\lambda_n - \phi))}{\cos(\phi)}\right]$$

This completely determines the solution. Taking any finite number of terms in the series produces an approximate solution exactly satisfying the differential equation and boundary conditions. Exact satisfaction of the zero initial condition would theoretically require an infinite number of series terms. However, using a 250-term series produces initial temperature values not exceeding 10^{-6}. Thus, the finite series is satisfactory for practical purposes.

The above equations were evaluated in a function called **heat**. The script file **slabheat** was also written to plot numerical results. The code and resulting Figures (Figures 9.3 and 9.4) appear below. This example clearly illustrates how well MATLAB handles complex arithmetic and complex valued functions.

TEMPERATURE VARIATION IN A SLAB

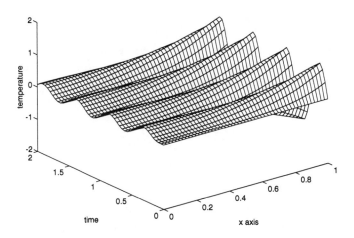

FIGURE 9.3. Temperature Variation in a Slab

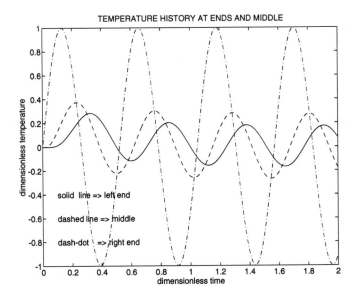

FIGURE 9.4. Temperature History at Ends and Middle

MATLAB EXAMPLE

9.3.0.10 Script File slabheat

```
 1: % Example: slabheat
 2: % ~~~~~~~~~~~~~~~~~~~
 3: % This program computes the temperature
 4: % variation in a one-dimensional slab with
 5: % the left end insulated and the right end
 6: % given a temperature variation sin(w*t).
 7: %
 8: % User m functions required:
 9: %    heat, removfil, genprint
10: %-----------------------------------------------
11:
12: [u1,t1,x1]=heat(12,0,2,50,0,1,51,250);
13: mesh(x1,t1,u1), axis([0 1 0 2 -2 2])
14: title('TEMPERATURE VARIATION IN A SLAB')
15: xlabel('x axis'); ylabel('time');
16: zlabel('temperature')
17: fprintf('\nPress RETURN to continue\n');
18: pause; genprint('tempsurf');
19:
20: [u2,t2,x2]=heat(12,0,2,150,0,1,3,250);
21: plot(t2,u2(:,1),'-w',t2,u2(:,2),'--w', ...
22:       t2,u2(:,3),'-.w')
23: title(['TEMPERATURE HISTORY AT ENDS' ...
24:         ' AND MIDDLE'])
25: xlabel('dimensionless time')
26: ylabel('dimensionless temperature')
27: text(.1,-.4,'solid  line => left end')
28: text(.1,-.6,'dashed line => middle')
29: text(.1,-.8,'dash-dot   => right end')
30: genprint('tempplot');
```

9.3.0.11 Function heat

```
 1: function [u,t,x]= ...
 2:           heat(w,tmin,tmax,nt,xmin,xmax,nx,nsum)
 3: %
 4: %[u,t,x]=heat(w,tmin,tmax,nt,xmin,xmax,nx,nsum)
 5: %~~~~~~~~~~~~~~~~~~~~~~~~~~~~~~~~~~~~~~~~~~~~~~~~
```

```
 6: % This function evaluates transient heat
 7: % conduction in a slab which has the left end
 8: % (x=0) insulated and has the right end (x=1)
 9: % subjected to a temperature variation
10: % sin(w*t). The initial temperature of the slab
11: % is zero.
12: %
13: % w             - frequency of the right side
14: %                 temperature variation
15: % tmin,tmax     - time limits for solution
16: % nt            - number of uniformly spaced
17: %                 time values used
18: % xmin,xmax     - position limits for solution.
19: %                 Values should lie between zero
20: %                 and one.
21: % nx            - number of equidistant x values
22: % nsum          - number of terms used in the
23: %                 series solution
24: % u             - matrix of temperature values.
25: %                 Time varies from row to row.
26: %                 x varies from column to column.
27: % t,x           - vectors of time and x values
28: %
29: % User m functions called:  none.
30: %----------------------------------------------
31:
32: t=tmin+(tmax-tmin)/(nt-1)*(0:nt-1);
33: x=xmin+(xmax-xmin)/(nx-1)*(0:nx-1)';
34: W=sqrt(-i*w); ln=pi*((1:nsum)-1/2);
35: v1=ln+W; v2=ln-W;
36: a=-imag((sin(v1)./v1+sin(v2)./v2)/cos(W));
37: u=imag(cos(W*x)*exp(i*w*t)/cos(W))+ ...
38:    (a(ones(nx,1),:).*cos(x*ln))* ...
39:    exp(-ln(:).^2*t);
40: u=u'; t=t(:);
```

FIGURE 9.5. Beam Geometry and Loading

9.4 Wave Propagation in a Beam with an Impact Moment Applied to One End

Analyzing the dynamic response caused when a time dependent moment acts on the end of an Euler beam involves a boundary value problem illustrating the use of Fourier series for solving linear partial differential equations. In the following example we consider a beam of uniform cross section which is pin-ended (hinged at the end) and is initially at rest. Suddenly, a harmonically varying moment $M_0 \cos(\Omega_0 T)$ is applied to the right end as shown in Figure 9.5. Determination of the resulting displacement and bending moment in the beam is desired.

Let U be transverse displacement, X longitudinal distance from the right end, and T time. The differential equation, boundary conditions and initial conditions characterizing the problem are

$$EI\frac{\partial^4 U}{\partial X^4} = -A\rho\frac{\partial^2 U}{\partial T^2} \qquad 0 < x < L \qquad T > 0$$

$$U(0,T) \;=\; 0 \qquad\qquad \frac{\partial^2 U}{\partial X^2}(0,T) \;=\; 0$$

$$U(L,T) \;=\; 0 \qquad\qquad \frac{\partial^2 U}{\partial X^2}(L,T) \;=\; M_0 \cos(\Omega_0 T)/(EI)$$

$$U(0,T) \;=\; 0 \qquad\qquad \frac{\partial U}{\partial T}(0,T) \;=\; 0$$

where L is beam length, EI is the product of elastic modulus and moment of inertia, and $A\rho$ is the product of cross section area and mass density.

This problem can be represented more conveniently by introducing dimensionless variables

x - dimensionless position $= X/L$

t - dimensionless time $= \left[\sqrt{(EI)/(A\rho)}\right] L^{-2}T$

u - dimensionless displacement $= (EI)/(M_0 L^2)U$

ω - dimensionless forcing frequency $= \left[\sqrt{(A\rho)/(EI)}\right] L^2 \Omega_0$

m - dimensionless bending moment $= \frac{\partial^2 u}{\partial x^2}$

The new boundary value problem is then

$$\frac{\partial^4 u}{\partial x^4} = -\frac{\partial^2 u}{\partial t^2} \qquad 0 < x < 1 \qquad t > 0$$

$$u(0,t) = 0 \qquad \frac{\partial^2 u}{\partial x^2}(0,t) = 0$$

$$u(1,t) = 0 \qquad \frac{\partial^2 u}{\partial x^2}(1,t) = 0$$

$$u(x,0) = 0 \qquad \frac{\partial u}{\partial t}(x,0) = 0 \qquad 0 < x < 1$$

The problem can be solved by combining a particular solution w which satisfies the differential equation and nonhomogeneous boundary conditions, with a homogeneous solution in series form which satisfies the differential equation and homogeneous boundary conditions. Thus we have $u = w + v$. The particular solution can be found in the form

$$u = f(x)\cos(\omega t)$$

where $f(x)$ satisfies
$$f''''(x) = \omega^2 f(x)$$
$$f(0) = f''(0) = f(1) \qquad f''(1) = 1$$

This ordinary differential equation is solvable as

$$f(x) = \sum_{k=1}^{4} c_k\, e^{s_k x}$$

where

$$s_k = \sqrt{\omega}\, e^{\pi\imath(k-1)/2} \qquad \imath = \sqrt{-1}$$

The boundary conditions require

$$\sum_{k=1}^{4} c_k = 0 \qquad \sum_{k=1}^{4} s_k^2 c_k = 1$$

$$\sum_{k=1}^{4} c_k \, e^{s_k} = 0 \qquad \sum_{k=1}^{4} c_k s_k^2 \, e^{s_k} = 0$$

Solving these simultaneous equations determines the particular solution. The particular solution satisfies initial conditions

$$w(0,t) = f(x) = \sum_{k=-\infty}^{\infty} c_k \, e^{\imath \pi k x} \qquad \frac{\partial w}{\partial t}(0,t) = 0$$

where $f(x)$ is expandable in a complex Fourier series as an odd valued function such that

$$f(x) = f(x+2) = -f(2-x) \qquad 0 < x < 1$$

This implies that $f(x)$ is represented as a sine series, or

$$f(x) = \sum_{k=1}^{\infty} a_k \sin(k\pi x)$$

with

$$a_k = -2 * \mathbf{imag}(c_k)$$

The homogeneous solution is representable as

$$v(x,t) = -\sum_{k=1}^{\infty} a_k \cos(\pi^2 k^2 t) \sin(k\pi x)$$

so that $w + v$ combine to satisfy the desired initial conditions of zero displacement and velocity.

Of course, perfect satisfaction of the initial conditions cannot be achieved without taking an infinite number of terms in the Fourier series. However, the series converges very rapidly because the coefficients are of order n^{-3}. When a hundred or more terms are used, an approximate solution produces results which satisfy the differential equation and boundary conditions, and which insignificantly violates the initial displacement condition. It is important to remember the nature of this error when examining the bending moment results presented below. Effects of high frequency components are very evident in the moment. Despite the oscillatory character of the moments, these results are exact for the initial displacement conditions produced by the truncated series. These displacements agree closely with the exact solution.

A program was written to evaluate the series solution to compute displacements and moments as functions of position and time. Plots and surfaces showing these quantities are presented along with timewise animations of the displacement and moment across the span. The computation involves the following steps:

1. evaluate $f(x)$

2. expand $f(x)$ using the FFT to get coefficients for the homogeneous series solution

3. combine the particular and homogeneous solution by summing the series for any number of terms desired by the user

4. plot u and m for selected times

5. plot surfaces showing $u(x,t)$ and $m(x,t)$

6. show animated plots of u and m

The principal parts of the program are shown in the table below.

rnbemimp	reads data and creates graphical output
beamresp	converts material property data to dimensionless form and calls **ndbemrsp**
ndbemrsp	construct the solution using Fourier series
sumser	sums the series for displacement and moment
animate	animates the time history of displacement and moment
removfil	deletes a file if it exists
genprint	saves file for printing

320

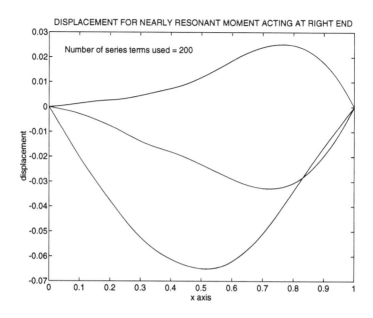

DISPLACEMENT FOR NEARLY RESONANT MOMENT ACTING AT RIGHT END

Number of series terms used = 200

FIGURE 9.6. Displacement Due to Impact Moment at Right End

The numerical results show the response for a beam subjected to a moment close to the natural frequencies of the beam. It can be shown that, in the dimensionless problem, the system of equations defining the particular solution becomes singular when ω assumes values of the form $k^2\pi^2$ for integer k. In that instance the series solution provided here will fail. However, ω values near to resonance can be used to show how the displacements and moments quickly become large. In our example we let EI, $A\rho$, l, and M_0 all equal unity, and $\omega = 0.95\pi^2$. Figures 9.6 and 9.7 show displacement and bending moment patterns shortly after motion is initiated. The surfaces in Figures 9.8 and 9.9 also show how the displacement and moment grow quickly with increasing time. The reader may find it interesting to run the program for various choices of ω and observe how dramatically the chosen forcing frequency affects results.

FIGURE 9.7. Bending Moment in the Beam

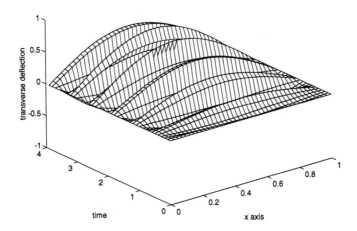

FIGURE 9.8. Displacement Growth Near Resonance

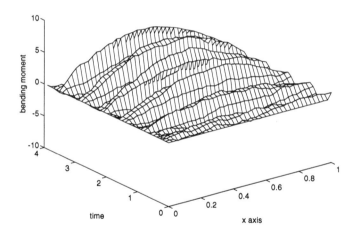

FIGURE 9.9. Moment Growth Near Resonance

MATLAB EXAMPLE

9.4.0.12   Script File rnbemimp

```
 1: % Example: rnbemimp
 2: % ~~~~~~~~~~~~~~~~~~~
 3: % This program analyzes an impact dynamics
 4: % problem for an elastic Euler beam of
 5: % constant cross section which is simply
 6: % supported at each end. The beam is initially
 7: % at rest when a harmonically varying moment
 8: % m0*cos(w0*t) is applied to the right end.
 9: % The resulting transverse displacements and
10: % bending moments are computed.  The
11: % displacement and moment are plotted as
12: % functions of x for the first several
13: % successive time steps. Animated plots of the
14: % entire displacement and moment history are
15: % also given.
16: %
17: % User m functions required:
18: %    beamresp, removfil, animate, sumser,
19: %    ndbemrsp, rnbemimp, genprint
20: %-----------------------------------------------
21:
22: fprintf('\nDYNAMICS OF A BEAM WITH AN ')
23: fprintf('OSCILLATORY END MOMENT\n');
24: ei=1; arho=1; len=1; m0=1; w0=.90*pi^2;
25: tmin=0; tmax=5; nt=101;
26: xmin=0; xmax=len; nx=81; ntrms=200;
27: [t,x,displ,mom]=beamresp(ei,arho,len,m0,w0,...
28:             tmin,tmax,nt,xmin,xmax,nx,ntrms);
29:
30: np=[3 5 8]; clf;
31: dip=displ(np,:); mop=mom(np,:);
32: plot(x,dip,'w');
33: xlabel('x axis'); ylabel('displacement')
34: hh=gca;
35: r(1:2)=get(hh,'XLim'); r(3:4)=get(hh,'YLim');
36: xp=r(1)+(r(2)-r(1))/10;
37: dp=r(3)+(r(4)-r(3))/10;
38: tstr=['DISPLACEMENT FOR NEARLY RESONANT' ...
39:       ' MOMENT ACTING AT RIGHT END'];
40: title(tstr);
```

```
41: text(xp,dp,['Number of series terms ' ...
42:             'used = ',int2str(ntrms)]);
43: fprintf('\nPress RETURN to continue'); pause;
44: genprint('3positns');
45:
46: clf; plot(x,mop,'w');
47: h=gca;
48: r(1:2)=get(h,'XLim'); r(3:4)=get(h,'YLim');
49: mp=r(3)+(r(4)-r(3))/10;
50: xlabel('x axis'); ylabel('moment');
51: tstr=['BENDING MOMENT FOR NEARLY RESONANT' ...
52:       ' MOMENT ACTING AT RIGHT END'];
53: title(tstr)
54: text(xp,mp,['Number of series terms ' ...
55:             'used = ',int2str(ntrms)]);
56: fprintf('\nPress RETURN to continue'); pause;
57: genprint('3moments');
58:
59: inct=2; incx=2;
60: ht=0.75; it=1:inct:.8*nt; ix=1:incx:nx;
61: tt=t(it); xx=x(ix);
62: dd=displ(it,ix); mm=mom(it,ix);
63: % a=mesh(xx,tt,dd);
64: % set(a,'edge','b'); set(a,'face','w')
65: a=mesh(xx,tt,dd);
66: tstr=['TRANSVERSE DEFLECTION AS A ' ...
67:       'FUNCTION OF TIME AND POSITION'];
68: title(tstr);
69: xlabel('x axis'); ylabel('time');
70: zlabel('transverse deflection')
71: fprintf('\nPress RETURN to continue'); pause;
72: genprint('dispsrf');
73:
74: % a=mesh(xx,tt,mm);
75: % set(a,'edge','b'); set(a,'face','w');
76: a=mesh(xx,tt,mm);
77: title(['BENDING MOMENT AS A FUNCTION ' ...
78:        'OF TIME AND POSITION'])
79: xlabel('x axis'); ylabel('time');
80: zlabel('bending moment')
81: fprintf('\nPress RETURN to animate ')
82: fprintf('beam deflection');
83: pause; genprint('momsrf');
84: animate(x,displ,.1,'TRANSVERSE DEFLECTION', ...
85:         'x axis','deflection');
```

```
86: fprintf('\nPress RETURN to animate ')
87: fprintf('bending moment\n');
88: pause;
89: animate(x,mom,.1,'BENDING MOMENT HISTORY', ...
90:         'x axis','moment');
91: disp('All Done')
```

9.4.0.13   Function beamresp

```
 1: function [t,x,displ,mom]= ...
 2:         beamresp(ei,arho,len,m0,w0,tmin,tmax, ...
 3:                  nt,xmin,xmax,nx,ntrms)
 4: %
 5: % [t,x,displ,mom]=beamresp(ei,arho,len,m0, ...
 6: %         w0,tmin,tmax,nt,xmin,xmax,nx,ntrms)
 7: % ~~~~~~~~~~~~~~~~~~~~~~~~~~~~~~~~~~~~~~~~~~~~
 8: % This function evaluates the time dependent
 9: % displacement and moment in a constant
10: % cross-section simply-supported beam which
11: % is initially at rest when a harmonically
12: % varying moment is suddenly applied at the
13: % right end.  The resulting time histories of
14: % displacement and moment are computed.
15: %
16: % ei        - modulus of elasticity times
17: %             moment of inertia
18: % arho      - mass per unit length of the
19: %             beam
20: % len       - beam length
21: % m0,w0     - amplitude and frequency of the
22: %             harmonically varying right end
23: %             moment
24: % tmin,tmax - minimum and maximum times for
25: %             the solution
26: % nt        - number of evenly spaced
27: %             solution times
28: % xmin,xmax - minimum and maximum position
29: %             coordinates for the solution.
30: %             These values should lie between
31: %             zero and len (x=0 and x=len at
32: %             the left and right ends).
33: % nx        - number of evenly spaced solution
34: %             positions
```

```
35: % ntrms      - number of terms used in the
36: %              Fourier sine series
37: % t          - vector of nt equally spaced time
38: %              values varying from tmin to tmax
39: % x          - vector of nx equally spaced
40: %              position values varying from
41: %              xmin to xmax
42: % displ      - matrix of transverse
43: %              displacements with time varying
44: %              from row to row, and position
45: %              varying from column to column
46: % mom        - matrix of bending moments with
47: %              time varying from row to row,
48: %              and position varying from column
49: %              to column
50: %
51: % User m functions called:  ndbemrsp
52: %---------------------------------------------
53:
54: tcof=sqrt(arho/ei)*len^2; dcof=m0*len^2/ei;
55: tmin=tmin/tcof; tmax=tmax/tcof; w=w0*tcof;
56: xmin=xmin/len; xmax=xmax/len;
57: [t,x,displ,mom]=...
58: ndbemrsp(w,tmin,tmax,nt,xmin,xmax,nx,ntrms);
59: t=t*tcof; x=x*len;
60: displ=displ*dcof; mom=mom*m0;
```

9.4.0.14   Function ndbemrsp

```
1: function [t,x,displ,mom]= ...
2:      ndbemrsp(w,tmin,tmax,nt,xmin,xmax,nx,ntrms)
3: %
4: % [t,x,displ,mom]=ndbemrsp(w,tmin,tmax,nt,...
5: %                       xmin,xmax,nx,ntrms)
6: % ~~~~~~~~~~~~~~~~~~~~~~~~~~~~~~~~~~~~~~~~~~~~~~
7: % This function evaluates the nondimensional
8: % displacement and moment in a constant
9: % cross-section simply-supported beam which
10: % is initially at rest when a harmonically
11: % varying moment of frequency w is suddenly
12: % applied at the right end. The resulting
13: % time history is computed.
14: %
```

```
15: % w              - frequency of the harmonically
16: %                  varying end moment
17: % tmin,tmax      - minimum and maximum
18: %                  dimensionless times
19: % nt             - number of evenly spaced
20: %                  solution times
21: % xmin,xmax      - minimum and maximum
22: %                  dimensionless position
23: %                  coordinates. These values
24: %                  should lie between zero and
25: %                  one (x=0 and x=1 give the
26: %                  left and right ends).
27: % nx             - number of evenly spaced
28: %                  solution positions
29: % ntrms          - number of terms used in the
30: %                  Fourier sine series
31: % t              - vector of nt equally spaced
32: %                  time values varying from
33: %                  tmin to tmax
34: % x              - vector of nx equally spaced
35: %                  position values varying
36: %                  from xmin to xmax
37: % displ          - matrix of dimensionless
38: %                  displacements with time
39: %                  varying from row to row,
40: %                  and position varying from
41: %                  column to column
42: % mom            - matrix of dimensionless
43: %                  bending moments with time
44: %                  varying from row to row, and
45: %                  position varying from column
46: %                  to column
47: %
48: % User m functions called:  sumser
49: %-------------------------------------------------
50:
51: if nargin < 8, w=0; end; nft=512; nh=nft/2;
52: xft=1/nh*(0:nh)';
53: x=xmin+(xmax-xmin)/(nx-1)*(0:nx-1)';
54: t=tmin+(tmax-tmin)/(nt-1)*(0:nt-1)';
55: cwt=cos(w*t);
56:
57: % Get particular solution for nonhomogeneous
58: % end condition
59: if w ==0 % Case for a constant end moment
```

```
60:    cp=[1 0 0 0; 0 0 2 0; 1 1 1 1; 0 0 2 6]\ ...
61:        [0;0;0;1];
62:    yp=[ones(x), x, x.^2, x.^3]*cp; yp=yp';
63:    mp=[zeros(nx,2), 2*ones(nx,1), 6*x]*cp;
64:    mp=mp';
65:    ypft=[ones(xft), xft, xft.^2, xft.^3]*cp;
66:
67: % Case where end moment oscillates
68: % with frequency w
69: else
70:    s=sqrt(w)*[1, i, -1, -i]; es=exp(s);
71:    cp=[ones(1,4); s.^2; es; es.*s.^2]\ ...
72:        [0; 0; 0; 1];
73:    yp=real(exp(x*s)*cp); yp=yp';
74:    mp=real(exp(x*s)*(cp.*s(:).^2)); mp=mp';
75:    ypft=real(exp(xft*s)*cp);
76: end
77:
78: % Fourier coefficients for
79: % particular solution
80: yft=-fft([ypft;-ypft(nh:-1:2)])/nft;
81:
82: % Sine series coefficients for
83: % homogeneous solution
84: acof=-2*imag(yft(2:ntrms+1));
85: ccof=pi*(1:ntrms)'; bcof=ccof.^2;
86:
87: % Sum series to evaluate Fourier
88: % series part of solution. Then combine
89: % with the particular solution.
90: displ=sumser(acof,bcof,ccof,'cos','sin',...
91:                     tmin,tmax,nt,xmin,xmax,nx);
92: displ=displ+cwt*yp; acof=acof.*bcof;
93: mom=sumser(acof,bcof,ccof,'cos','sin',...
94:                     tmin,tmax,nt,xmin,xmax,nx);
95: mom=-mom+cwt*mp;
```

9.4.0.15   Function sumser

```
1: function [u,t,x] = sumser(a,b,c,funt,funx, ...
2:                     tmin,tmax,nt,xmin,xmax,nx)
3: %
4: % [u,t,x] = sumser(a,b,c,funt,funx,tmin, ...
```

```
 5: %                          tmax,nt,xmin,xmax,nx)
 6: % ~~~~~~~~~~~~~~~~~~~~~~~~~~~~~~~~~~~~~~~~~~~~~~~~
 7: % This function evaluates a function U(t,x)
 8: % which is defined by a finite series. The
 9: % series is evaluated for t and x values taken
10: % on a rectangular grid network. The matrix u
11: % has elements specified by the following
12: % series summation:
13: %
14: % u(i,j)  =   sum(    a(k)*funt(t(i)* ...
15: %           k=1:nsum
16: %                     b(k))*funx(c(k)*x(j))
17: %
18: % where nsum is the length of each of the
19: % vectors a, b, and c.
20: %
21: % a,b,c         - vectors of coefficients in
22: %                 the series
23: % funct,funx    - functions which accept a
24: %                 matrix argument. funct is
25: %                 evaluated for an argument of
26: %                 the form func(t*b) where t is
27: %                 a column and b is a row. funx
28: %                 is evaluated for an argument
29: %                 of the form funx(c*x) where
30: %                 c is a column and x is a row.
31: % tmin,tmax,nt - produces vector t with nt
32: %                 evenly spaced values between
33: %                 tmin and tmax
34: % xmin,xmax,nx - produces vector x with nx
35: %                 evenly spaced values between
36: %                 xmin and xmax
37: % u             - the nt by nx matrix
38: %                 containing values of the
39: %                 series evaluated at t(i),x(j),
40: %                 for i=1:nt and j=1:nx
41: % t,x           - column vectors containing t
42: %                 and x values. These output
43: %                 values are optional.
44: %
45: % User m functions called:  none.
46: %-------------------------------------------------
47:
48: tt=(tmin:(tmax-tmin)/(nt-1):tmax)';
49: xx=(xmin:(xmax-xmin)/(nx-1):xmax); a=a(:).';
```

```
50: u=a(ones(nt,1),:).*feval(funt,tt*b(:).')*...
51: feval(funx,c(:)*xx);
52: if nargout>1, t=tt; x=xx'; end
```

9.4.0.16   Function animate

```
 1: function animate(x,u,tpause,titl,xlabl,ylabl)
 2: %
 3: % animate(x,u,tpause,titl,xlabl,ylabl)
 4: % ~~~~~~~~~~~~~~~~~~~~~~~~~~~~~~~~~~~~~~~~~
 5: % This function draws an animated plot of data
 6: % values stored in array u.  The different
 7: % columns of u correspond to position values
 8: % in vector x.  The successive rows of u
 9: % correspond to different times. Parameter
10: % tpause controls the speed of animation.
11: %
12: % u       - matrix of values to animate plots
13: %             of u versus x
14: % x       - spatial positions for different
15: %             columns of u
16: % tpause  - clock seconds between output of
17: %             frames. The default is .1 secs
18: %             when tpause is left out. When
19: %             tpause=0, a new frame appears
20: %             when the user presses any key.
21: % titl    - graph title
22: % xlabl   - label for horizontal axis
23: % ylabl   - label for vertical axis
24: %
25: % User m functions called:  none
26: %-------------------------------------------------
27:
28: clf;
29: if nargin<6, ylabl=''; end;
30: if nargin<5, xlabl=''; end
31: if nargin<4, titl=''; end;
32: if nargin<3, tpause=.1; end
33:
34: [ntime,nxpts]=size(u);
35: umin=min(u(:)); umax=max(u(:));
36: udif=umax-umin;
37: uavg=.5*(umin+umax);
```

```
38: xmin=min(x); xmax=max(x);
39: xdif=xmax-xmin; xavg=.5*(xmin+xmax);
40: xwmin=xavg-.55*xdif; xwmax=xavg+.55*xdif;
41: uwmin=uavg-.55*udif; uwmax=uavg+.55*udif; clf
42: axis([xwmin,xwmax,uwmin,uwmax]); title(titl)
43: xlabel(xlabl); ylabel(ylabl); hold on
44:
45: for j=1:ntime
46:   ut=u(j,:); plot(x,ut,'w');
47:   if tpause==0
48:     pause;
49:   else
50:     pause(tpause);
51:   end
52:   if j==ntime, break, else, cla; end
53: end
54: genprint('cntltrac.ps')
55:
56: hold off; clf;
```

9.5 Torsional Stresses in a Beam of Rectangular Cross Section

Elastic beams of uniform cross section are commonly used structural members. Evaluation of the stresses caused when beams undergo torsional moments depends on finding a particular type of complex valued function. This function is analytic inside the beam cross section and has its imaginary part known on the boundary [62]. The shear stresses τ_x and τ_y are obtained from the stress function $f(z)$ of the complex variable $z = x + iy$ according to

$$\frac{\tau_{zx} - i\tau_{zy}}{\mu\alpha} = f'(z) - i\bar{z}$$

where μ is the shear modulus and α is the twist per unit length. In the case for a simply connected cross section, such as a rectangle or a semicircle, the necessary boundary condition is

$$\mathbf{imag}[f(z)] = \frac{1}{2}|z|^2$$

at all boundary points. It can also be shown that the torsional moment causes the beam cross section to warp. The warped shape is given by the real part of $f(z)$.

The geometry we will analyze is rectangular. As long as the ratio of side length remains fairly close to unity, $f(z)$ can be approximated well in series form as

$$f(z) = i\sum_{j=1}^{n} c_j \left(\frac{z}{s}\right)^{2j-2}$$

where $c_1, \ldots, c_n$ are real coefficients computed to satisfy the boundary conditions in the least square sense. Parameter s is used for scaling to prevent occurrence of large numbers when n becomes large. We take a rectangle with sides parallel to the coordinate axes and assume side lengths of $2a$ and $2b$ for the horizontal and vertical directions, respectively. The scaling parameter will be chosen as the larger of a and b. The boundary conditions state that for any point z_i on the boundary we should have

$$\sum_{j=1}^{n} c_j \, \mathbf{real}\left[\left(\frac{z_i}{s}\right)^{2j-2}\right] = \frac{1}{2}|z_i|^2$$

WARPING OF THE CROSS SECTION

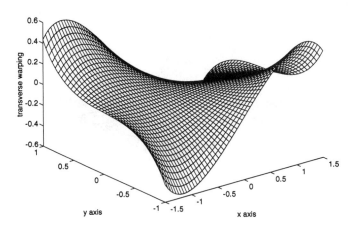

FIGURE 9.10. Warping of the Cross Section

Once the series coefficients are found, then shear stresses are computed as

$$\frac{\tau_x - \imath \tau_y}{\mu \alpha} = -\imath \bar{z} + 2\imath z^{-1} \sum_{j=2}^{n} (j-1) c_j \left(\frac{z}{s}\right)^{2j-3}$$

A program was written to compute stresses in a rectangular beam and to show graphically the cross section warping and the dimensionless stress values. The program is short and the necessary calculations are almost self explanatory. It is worthwhile to observe, however, the ease with which MATLAB handles complex functions. Note how intrinsic function **linspace** is used to generate boundary data and **meshdom** is used to generate a grid of complex values (see lines 50, 51, 71, 72, and 73 of function **recstrs**). The sample problem employs a rectangle of dimension 2 units by 4 units. The maximum stress occurs at the middle of the longest side. Figures 9.10–9.13 plot the results of this analysis.

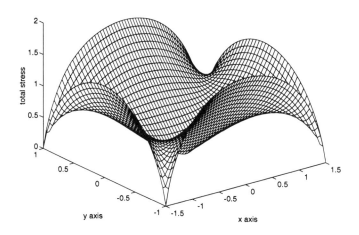

FIGURE 9.11. Total Stress Surface

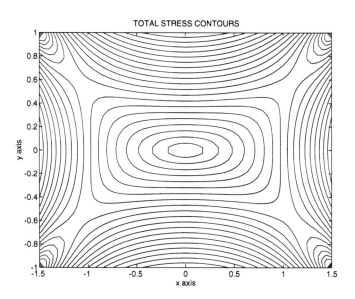

FIGURE 9.12. Total Stress Contours

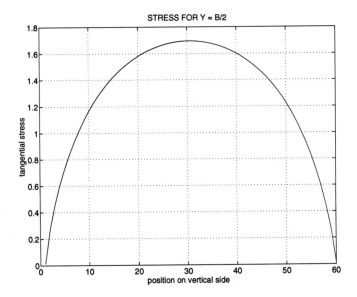

FIGURE 9.13. Stress For $Y = B/2$

MATLAB EXAMPLE

9.5.0.17 Output from Example

```
>> rector

=== TORSIONAL STRESS CALCULATION IN A RECTANGULAR ===
===      BEAM USING LEAST SQUARE APPROXIMATION      ===

Input the lengths of the horizontal and
the vertical sides
 ? 3,2

To approximate boundary conditions, give the number
of least square points on the horizontal side and the
number of least square points on the vertical side
(40,40 is typical for a square)
 ? 60,40

Input the number of terms used in the stress function
(20 terms is usually enough)
 ? 30

To display the stress and displacement quantities on
a grid of the cross section, input the number of grid
lines for the X direction and for the Y direction.
(25,25 is usually adequate)
 ? 60,40
```

9.5.0.18   Script File rector

```
 1: % Example:   rector
 2: % ~~~~~~~~~~~~~~~~~
 3: % This program employs point matching to obtain
 4: % an approximate solution for torsional
 5: % stresses in a Saint Venant beam of
 6: % rectangular cross section. The complex stress
 7: % function is analytic inside the rectangle and
 8: % has its real part equal to abs(z*z)/2 on the
 9: % boundary. The problem is solved approximately
10: % using a polynomial stress function which fits
11: % the boundary condition in the least square
12: % sense. Surfaces and contour curves describing
13: % the stress and deformation pattern in the
14: % beam cross section are drawn.
15: %
16: % User m functions required:
17: %    recstrs, read, genprint, removfil
18: %-------------------------------------------------
19:
20: clear;
21: fprintf('\n===   TORSIONAL STRESS CALCULATION');
22: fprintf(' IN A RECTANGULAR  ===');
23: fprintf('\n===       BEAM USING LEAST SQUARE ');
24: fprintf('APPROXIMATION       ===\n')
25: fprintf('\nInput the lengths of the ')
26: fprintf('horizontal and the vertical sides');
27: [a,b]=read; a=a/2; b=b/2;
28: fprintf('\nTo approximate the boundary ')
29: fprintf('conditions, give the number of');
30: fprintf('\nleast square points on the ')
31: fprintf('horizontal side and the number of');
32: fprintf('\nleast square points on the ')
33: fprintf('vertical side');
34: fprintf('\n(40,40 is typical for a square)');
35: [nsega,nsegb]=read;
36: nsega=fix(nsega/2); nsegb=fix(nsegb/2);
37: fprintf('\nInput the number of terms ')
38: fprintf('used in the stress function')
39: fprintf('\n(20 terms is usually enough)');
40: ntrms=input('? ');
41: fprintf('\nTo display the stress and ')
42: fprintf('displacement quantities on')
```

```
43: fprintf('\na grid of the cross section,')
44: fprintf(' input the number of grid')
45: fprintf('\nlines for the X direction and ')
46: fprintf('for the Y direction.')
47: fprintf('\n(25,25 is usually adequate)')
48: [nxout,nyout]=read;
49:
50: [c,phi,stres,z] = ...
51:    recstrs(a,nsega,b,nsegb,ntrms,nxout,nyout);
52: disp(''); disp('All Done')
```

9.5.0.19   Function recstrs

```
 1: function [c,phi,stres,z]=...
 2:    recstrs(a,nsega,b,nsegb,ntrms,nxout,nyout)
 3: %
 4: % [c,phi,stres,z]=...
 5: %    recstrs(a,nsega,b,nsegb,ntrms,nxout,nyout)
 6: % ~~~~~~~~~~~~~~~~~~~~~~~~~~~~~~~~~~~~~~~~~~~~~~~~
 7: % This function employs point matching to
 8: % obtain an approximate solution for torsional
 9: % stresses in a Saint Venant beam of
10: % rectangular cross section. The complex stress
11: % function is analytic inside the rectangle
12: % and has its real part equal to abs(z*z)/2 on
13: % the boundary. The problem is solved
14: % approximately using a polynomial stress
15: % function which fits the boundary condition
16: % in the least square sense. The beam is 2*a
17: % wide along the x axis and 2*b deep along
18: % the y axis. The shear stresses in the beam
19: % are given by the complex stress formula:
20: %
21: % (tauzx-i*tauzy)/(mu*alpha) = -i*conj(z)+f'(z)
22: %
23: % where
24: %
25: %    f(z)=i*sum( c(j)*z^(2*j-2), j=1:ntrms )
26: %
27: % and c(j) are real.
28: %
29: % a,b     - half the side lengths of the
30: %                horizontal and vertical sides
```

```
31: % nsega,   - numbers of subintervals used in
32: % nsegb        formation of least square
33: %               equations
34: % ntrms    - number of polynomial terms used in
35: %               polynomial stress function
36: % nxout,   - number of grid points used to
37: % nyout        evaluate output
38: % c        - coefficeints defining the stress
39: %               function
40: % phi      - values of the membrane function
41: % stres    - array of complex stress values
42: % z        - complex point array at which
43: %               stresses are found
44: %
45: % User m functions called:  genprint
46: %-----------------------------------------------
47:
48: % Generate vector zbdry of boundary points
49: % for point matching.
50: db=b/nsegb; zvert=linspace(a,a+i*(b-db),nsegb);
51: zhoriz=linspace(a+i*b,i*b,nsega+1);
52: zbdry=[zvert(:);zhoriz(:)];
53:
54: % Determine a scaling parameter used to
55: % prevent occurrence of large numbers when
56: % high powers of z are used
57: s=max(a,b);
58:
59: % Form the least square equations to impose
60: % the boundary conditions.
61: neq=length(zbdry); amat=ones(neq,ntrms);
62: ztmp=(zbdry/s).^2; bvec=.5*abs(zbdry).^2;
63: for j=2:ntrms
64:   amat(:,j)=amat(:,j-1).*ztmp; end
65:
66: % Solve the least square equations.
67: amat=real(amat); c=pinv(amat)*bvec;
68:
69: % Generate grid points to evaluate
70: % the solution.
71: xsid=linspace(-a,a,nxout);
72: ysid=linspace(-b,b,nyout);
73: [xg,yg]=meshdom(xsid,ysid);
74: z=xg+i*yg; zz=(z/s).^2;
75:
```

```
76: % Evaluate the warping function
77: phi=-imag(polyval(flipud(c),zz));
78:
79: % Evaluate stresses and plot results
80: cc=(2*(1:ntrms)-2)'.*c;
81: stres=-i*conj(z)+i* ...
82:       polyval(flipud(cc),zz)./(z+eps*(z==0));
83: am=num2str(-a);ap=num2str(a);
84: bm=num2str(-b);bp=num2str(b);
85:
86: % Plot results
87: fprintf('\nPress RETURN to plot ')
88: fprintf('warping surface');
89: pause; hold off; clf; mesh(xg,yg,phi);
90: title('WARPING OF THE CROSS SECTION')
91: xlabel('x axis'); ylabel('y axis');
92: zlabel('transverse warping');
93: fprintf('\nPress RETURN to show total ')
94: fprintf('stress surface');
95: pause; genprint('warpsurf');
96: mesh(xg,yg,abs(stres));
97: title('TOTAL STRESS SURFACE')
98: xlabel('x axis'); ylabel('y axis');
99: zlabel('total stress');
100: fprintf('\nPress RETURN to display ')
101: fprintf('stress contours'); pause
102: genprint('torstrsu')
103: contour(xg,yg,abs(stres),20);
104: title('TOTAL STRESS CONTOURS')
105: xlabel('x axis'); ylabel('y axis');
106: fprintf('\nPress RETURN to plot');
107: fprintf(' maximum stress on rectangle side\n')
108: pause; genprint('torcontu')
109: plot(abs(stres(1,:)))
110: grid; ylabel('tangential stress');
111: xlabel('position on vertical side')
112: title('STRESS FOR Y = B/2');
113: genprint('torstsid');
```

9.6   Accuracy Comparison for Euler Beam Natural Frequencies Obtained by Finite Element and Finite Difference Methods

This chapter concludes with an example involving natural frequency computation for a cantilever beam. Accuracy comparisons are made among results from the following three methods a) solution of the frequency equation for the true continuum model, b) approximation of the equations of motion using finite difference to replace spatial derivatives, and c) use of finite element methods implying a piecewise cubic spatial interpolation of the displacement field. The first method is less appealing as a general first method tool than the last two methods because the frequency equation would be awkward to obtain for geometries of variable cross section. Frequencies found using finite difference and finite element methods are compared with results from the exact model; and it is observed that the finite element method produces results which are superior to those from finite differences for comparable degrees-of-freedom. In addition, the natural frequencies and mode shapes given by finite elements are used to compute and animate system response produced when a beam, initially at rest, is suddenly subjected to two concentrated loads.

9.6.1   MATHEMATICAL FORMULATION

The differential equation governing transverse vibrations of an elastic beam of constant depth is [59]

$$EI\frac{\partial^4 Y}{\partial X^4} = -\rho\frac{\partial^2 Y}{\partial T^2} + W(X,T) \qquad 0 \le X \le \ell \qquad T \ge 0$$

where

$$
\begin{aligned}
Y(X,T) &\;-\; \text{transverse displacement} \\
X &\;-\; \text{horizontal position along the beam length} \\
T &\;-\; \text{time} \\
EI &\;-\; \text{product of moment of inertia and Young's} \\
& \qquad \text{modulus} \\
\rho &\;-\; \text{mass per unit length of the beam} \\
W(X,T) &\;-\; \text{external applied force per unit length}
\end{aligned}
$$

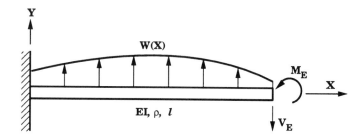

FIGURE 9.14. Cantilever Beam Subjected to Impact Loading

In the present study, we consider the cantilever beam shown in Figure 9.14, having end conditions which are

$$Y(0,T) = 0 \qquad \frac{\partial Y(0,T)}{\partial X} = 0$$

and

$$EI\frac{\partial^2 Y(\ell,T)}{\partial X^2} = M_E(T) \qquad EI\frac{\partial^3 Y(\ell,T)}{\partial X^3} = V_E(T)$$

This problem can be expressed more concisely using dimensionless variables

$$x = \frac{X}{\ell} \qquad y = \frac{Y}{\ell} \qquad t = \sqrt{\frac{EI}{\rho}}\left(\frac{T}{\ell^2}\right)$$

Then the differential equations and boundary conditions become

$$\frac{\partial^4 y}{\partial x^4} = -\frac{\partial^2 y}{\partial t^2} + w(x,t) \qquad 0 \le x \le 1 \qquad t \ge 0$$

$$y(0,t) = 0 \qquad \frac{\partial y}{\partial x}(0,t) = 0$$

$$\frac{\partial^2 y}{\partial x^2}(1,t) = m_e(t) \qquad \frac{\partial^3 y}{\partial x^3}(1,t) = v_e(t)$$

where

$$w = (W\ell^3)/(EI) \qquad m_e = (M_E\ell)/(EI)$$

and

$$v_e = (V_E \ell^2)/(EI)$$

The natural frequencies of the system are obtained by computing homogeneous solutions of the form $y(x,t) = f(x)\sin(\omega t)$ which exist when $w = m_e = v_e = 0$. This implies

$$\frac{d^4 f}{dx^4} = \lambda^4 f \qquad \lambda = \sqrt{\omega}$$

subject to

$$f(0) = 0 \qquad f'(0) = 0 \qquad f''(1) = 0 \qquad f'''(1) = 0$$

The solution satisfying this fourth order differential equation with homogeneous boundary conditions has the form

$$\begin{aligned} f \;=\; & [\cos(\lambda x) - \cosh(\lambda x)][\sin(\lambda) + \sinh(\lambda)] - \\ & [\sin(\lambda x) - \sinh(\lambda x)][\cos(\lambda) + \cosh(\lambda)] \end{aligned}$$

where λ values must be roots of the frequency equation

$$p(\lambda) = \cos(\lambda) + 1/\cosh(\lambda) = 0$$

Although the roots cannot be obtained explicitly, asymptotic approximations valid for large n are evidently

$$\lambda_n = (2k - 1)\pi/2$$

These estimates can be used as the starting points for a rapidly convergent Newton iteration indicated by

$$\lambda_{NEW} = \lambda_{OLD} - p(\lambda_{OLD})/p'(\lambda_{OLD})$$

with

$$p'(\lambda) = -\sin(\lambda) - [\sinh(\lambda)/\cosh^2(\lambda)]$$

The exact solution will be used to compare related results produced by finite difference and finite element methods. First we consider finite differences. The following difference formulas having a quadratic truncation error derivable from Taylor's series [1]:

$$
\begin{aligned}
\frac{\partial^4 y}{\partial x^4}(x) &= [y(x-2h) - 4y(x-h) + 6y(x) - \\
&\quad 4y(x+h) + y(x+2h)]/h^4 \\
y'(x) &= [-y(x-h) + y(x+h)]/(2h) \\
y''(x) &= [y(x-h) - 2y(x) + y(x+h)]/h^2 \\
y'''(x) &= [-y(x-2h) + 2y(x-h) - 2y(x+h) + \\
&\quad y(x+2h)]/(2h^3)
\end{aligned}
$$

The step-size will be taken as $h = 1/n$ so $x_j = jh$, $0 \le j \le n$. Therefore, x_0 is at the left end and x_n is at the right end of the beam. It is desirable to include additional fictitious points x_{-1}, x_{n+1}, and x_{n+2}. Then the left end conditions imply

$$
y_0 = y_1 \qquad \text{and} \qquad y_{-1} = y_1
$$

and the right end conditions imply

$$
y_{n+1} = -y_{n-1} + 2y_n
$$

$$
y_{n+2} = y_{n-2} - 4y_{n-1} + 4y_n
$$

Using these relations, the algebraic eigenvalue problem derived from the difference approximation is

$$
\begin{aligned}
7y_1 - 4y_2 + y_3 &= \tilde{\lambda} y_1 & \tilde{\lambda} = h^4 \lambda \\
-4y_1 + 6y_2 - 4y_3 + y_4 &= \tilde{\lambda} y_2 \\
y_{j-2} - 4y_{j-1} + 6y_j - 4y_{j+1} + y_{j+2} &= \tilde{\lambda} y_j & 2 < j < (n-1) \\
y_{n-3} - 4y_{n-2} + 5y_{n-1} - 2y_n &= \tilde{\lambda} y_{n-1} \\
2y_{n-2} - 4y_{n-1} + 2y_n &= \tilde{\lambda} y_n
\end{aligned}
$$

This can be solved for the natural frequencies provided satisfactory eigenvalue calculation software is available.

The finite element method leads to a similar problem involving global mass and stiffness matrices [46]. When we consider a single beam element of mass m and length ℓ, the elemental mass and stiffness matrices found using a cubically varying displacement approximation are

$$
M_e = \frac{m}{420}
\begin{bmatrix}
156 & 22\ell & 54 & -13\ell \\
22\ell & 4\ell^2 & 13\ell & -3\ell^2 \\
54 & 13\ell & 156 & -22\ell \\
-13\ell & -3\ell^2 & -22\ell & 4\ell^2
\end{bmatrix}
$$

$$
K_e = \frac{EI}{\ell^3}
\begin{bmatrix}
6 & 3\ell & -6 & 3\ell \\
3\ell & 2\ell^2 & -3\ell & \ell^2 \\
-6 & -3\ell & 6 & -3\ell \\
3\ell & \ell^2 & -3\ell & 2\ell^2
\end{bmatrix}
$$

and the elemental equation of motion has the form

$$
M_e Y_e'' + K_e Y_e = F_e
$$

where

$$
Y_e = [Y_1, Y_1', Y_2, Y_2']^T
$$

$$
F_e = [F_1, M_1, F_2, M_2]^T
$$

are generalized elemental displacement and force vectors. The global equation of motion is obtained as an assembly of element matrices and has the form

$$
MY'' + KY = F
$$

A system with N elements involves $N + 1$ nodal points. For the cantilever beam studied here both Y_0 and Y_0' are zero, so removing these two variables leaves a system of $n = 2N$ unknowns. The solution of this equation in the case of a non-resonant harmonic forcing function will be discussed further. The matrix analog of the simple harmonic equation is

$$
M\ddot{Y} + KY = F_1 \cos(\omega t) + F_2 \sin(\omega t)
$$

with initial conditions

$$
Y(0) = Y_0 \qquad \dot{Y}(0) = V_0
$$

Solution of this differential equation is facilitated by using a particular solution and a homogeneous solution

$$Y = Y_P + Y_H$$

where

$$Y_H = Y_1 \cos(\omega t) + Y_2 \sin(\omega t)$$

with

$$Y_j = (K - \omega^2 M)^{-1} F_j \qquad j = 1, 2$$

This assumes, of course, that $K - \omega^2 M$ is nonsingular. The homogeneous equation satisfies

$$M \ddot{Y}_H + K Y_H = 0$$

with the initial conditions

$$Y_H(0) = Y_0 - Y_1 \qquad \dot{Y}_H(0) = V_0 - \omega Y_2$$

The homogeneous solution components have the form

$$Y_{jH} = U_j \cos(\omega_j t + \phi_j)$$

where ω_j and U_j are natural frequencies and modal vectors satisfying the eigenvalue equation

$$K U_j = \omega_j^2 M U_j$$

Consequently, the homogeneous solution completing the modal response is

$$Y_H(t) = \sum_{j=1}^{n} U_j [\cos(\omega_j t) c_j + \sin(\omega_j t) d_j / \omega_j]$$

where c_j and d_j are computed to satisfy the initial conditions which require

$$C = U^{-1}(Y_0 - Y_1) \qquad D = U^{-1}(V_0 - \omega Y_2)$$

The next section presents the MATLAB program. Natural frequencies from finite difference and finite element matrices are compared and modal vectors from the finite element method are used to analyze a time response problem.

9.6.2 DISCUSSION OF THE CODE

A program was written to compare exact frequencies from the original continuous beam model with approximations produced using finite differences and finite elements. The finite element results were also employed to calculate a time response by modal superposition for any structure which has general mass and stiffness matrices, and is subjected to loads which are constant or harmonically varying.

The code below is longer than earlier problem solutions because various MATLAB capabilities are applied to three different solution methods. The following function summary involves eight segments, several of which were used earlier in the text.

cbfreq	driver to input data, call computation modules, and print results
cbfrqnwm	function to compute exact natural frequencies by Newton's method for root calculation
cbfrqfdm	forms equations of motion using finite difference and calls **eig** to compute natural frequencies
cbfrqfem	uses the finite element method to form the equation of motion and calls **eig** to compute natural frequencies and modal vectors
frud	function which solves the structural dynamics equation by methods developed in Chapter 7
examplmo	evaluates the response caused when a downward load at the middle and an upward load at the free end are applied
animate	plots successive positions of the beam to animate the motion
plotsave	plots the beam frequencies for the three methods. Also plots percent errors showing how accurate finite element and finite difference methods are
read	reads a sequence of numbers

Several characteristics of the functions assembled for this pro-

gram are worth examining in detail. The next table contains re-
marks relevant to the code.

Routine	Line	Operation
Output		Natural frequencies are printed along with error percentages. The output shown here has been extracted from the actual output to show only the highest and lowest frequencies.
cbfrqnwm	21	Asymptotic estimates are used to start a Newton method iteration.
	26-32	Root corrections are carried out for all roots until the correction to any root is sufficiently small.
cbfrqfdm	23-24	The equations of motion are formed without corrections for end conditions.
	26-33	End conditions are applied.
	37	**eig** computes the frequencies.
cbfrqfem	28-33	Form elemental mass matrix.
	35-39	Form elemental stiffness matrix.
	44-48	Global equations of motion are formed using an element by element loop.
	52	Boundary conditions are applied requiring zero displacement and slope at the left end, and zero moment and shear at the right end.
	54-61	Frequencies and modal vectors are computed. Note that modal vector computation is made optional since this takes longer than only computing frequencies.
		continued on next page

Routine	Line	Operation
		continued from previous page
frud		Compute time response by modal superposition. Theoretical details pertaining to this function appear in Chapter 7.
examplmo	33	The time step and maximum time for response calculation is selected.
	35-36	Function **frud** is used to compute displacement and rotation response. Only displacement is saved.
	39-44	Free end displacement is plotted.
	46-53	A surface showing displacement as a function of position and time is shown.
	56	Function **animate** is called.
animate	36-42	Window limits are determined.
	46-54	Each position is plotted. Then it is made invisible before proceeding to the next position.
plotsave		Plot and save graphs showing the frequencies and error percentages.

TABLE 9.1. Description of Code in Example

9.6.3 NUMERICAL RESULTS

The dimensionless frequency estimates from finite difference and finite element methods were compared for various numbers of degrees-of-freedom. Typical program output for $n = 100$ is shown at the end of this section. The frequency results and error percentages are shown in Figures 9.15 and 9.16. It is evident that the finite difference frequencies are consistently low and the finite element results are consistently high. The finite difference estimates degrade smoothly with increasing order. The finite element frequencies are surprisingly accurate for ω_k when $k < n/2$. At $k = n/2$ and $k = n$, the finite element error jumps sharply. This peculiar error jump halfway through the spectrum has also been observed in [46]. The most important and useful result seen from Figure 9.16 is that in order to obtain a particular number of fre-

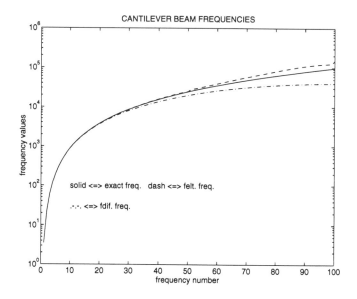

FIGURE 9.15. Cantilever Beam Frequencies

quencies, say N, which are accurate within 3.5%, it is necessary to employ more than $2N$ elements and keep only half of the predicted values.

The final result presented is the time response of a beam which is initially at rest when a concentrated downward load of five units is applied at the middle and a one unit upward load is applied at the free end. The time history was computed using function **frud**. Figure 9.17 shows the time history of the free end. Figure 9.18 is a surface plot illustrating how the deflection pattern changes with time. Finally, Figure 9.19 shows successive deflection positions produced by function **animate**. The output was obtained by suppressing the graph clearing option for successive configurations.

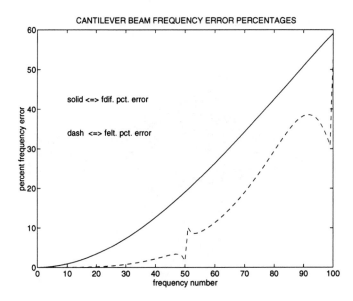

FIGURE 9.16. Cantilever Beam Frequency Error Percentages

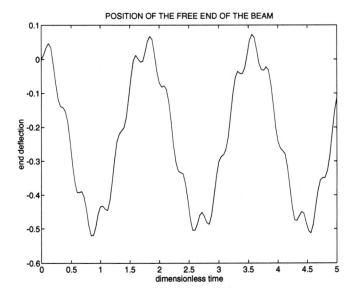

FIGURE 9.17. Position of the Free End of the Beam

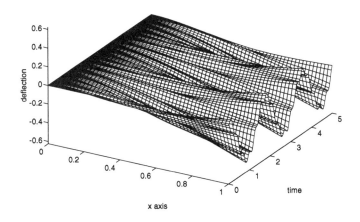

FIGURE 9.18. Beam Deflection History for Varying Time

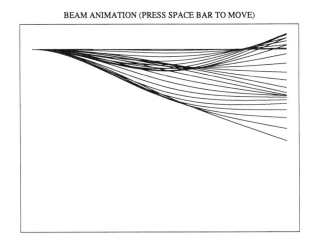

FIGURE 9.19. Beam Animation

MATLAB EXAMPLE

9.6.3.1   Output from Example

```
>> cbfreq
```

```
CANTILEVER BEAM FREQUENCIES BY FINITE DIFFERENCE AND
          FINITE ELEMENT APPROXIMATION
```

```
Give the number of frequencies to be computed
(use an even number greater than 2)
? > 100
```

freq. number	exact. freq.	fdif. freq.	fd. pct. error	felt. freq.	fe. pct. error
1	3.51602e+00	3.51572e+00	-0.008	3.51602e+00	0.000
2	2.20345e+01	2.20250e+01	-0.043	2.20345e+01	0.000
3	6.16972e+01	6.16414e+01	-0.090	6.16972e+01	0.000
4	1.20902e+02	1.20714e+02	-0.155	1.20902e+02	0.000
5	1.99860e+02	1.99386e+02	-0.237	1.99860e+02	0.000
6	2.98556e+02	2.97558e+02	-0.334	2.98558e+02	0.001
7	4.16991e+02	4.15123e+02	-0.448	4.16999e+02	0.002
8	5.55165e+02	5.51957e+02	-0.578	5.55184e+02	0.003
9	7.13079e+02	7.07918e+02	-0.724	7.13119e+02	0.006
10	8.90732e+02	8.82842e+02	-0.886	8.90809e+02	0.009
11	1.08812e+03	1.07655e+03	-1.064	1.08826e+03	0.013
12	1.30526e+03	1.28884e+03	-1.257	1.30550e+03	0.019
13	1.54213e+03	1.51950e+03	-1.467	1.54252e+03	0.026
14	1.79874e+03	1.76830e+03	-1.692	1.79937e+03	0.035
15	2.07508e+03	2.03497e+03	-1.933	2.07605e+03	0.047
16	2.37117e+03	2.31926e+03	-2.189	2.37261e+03	0.061
17	2.68700e+03	2.62088e+03	-2.461	2.68908e+03	0.077
18	3.02257e+03	2.93951e+03	-2.748	3.02551e+03	0.098
19	3.37787e+03	3.27486e+03	-3.050	3.38197e+03	0.121
20	3.75292e+03	3.62657e+03	-3.367	3.75851e+03	0.149

```
       ======  INTERMEDIATE LINES OF OUTPUT DELETED  ======
```

90	7.90580e+04	3.88340e+04	-50.879	1.09328e+05	38.288
91	8.08345e+04	3.90347e+04	-51.710	1.11989e+05	38.541
92	8.26308e+04	3.92169e+04	-52.540	1.14512e+05	38.582
93	8.44468e+04	3.93804e+04	-53.367	1.16860e+05	38.384
94	8.62825e+04	3.95250e+04	-54.191	1.18999e+05	37.917
95	8.81380e+04	3.96507e+04	-55.013	1.20889e+05	37.159
96	9.00133e+04	3.97572e+04	-55.832	1.22496e+05	36.086
97	9.19082e+04	3.98445e+04	-56.648	1.23786e+05	34.684
98	9.38229e+04	3.99125e+04	-57.460	1.24730e+05	32.941
99	9.57574e+04	3.99611e+04	-58.268	1.25305e+05	30.857
100	9.77116e+04	3.99903e+04	-59.073	1.49694e+05	53.200

Evaluate the time response resulting from a
concentrated downward load at the middle and
an upward end load.

input the time step and the maximum time
(0.04 and 5.0) are typical. Use 0,0 to stop

? .04,5

Evaluate the time response resulting from a
concentrated downward load at the middle and
an upward end load.

input the time step and the maximum time
(0.04 and 5.0) are typical. Use 0,0 to stop

? 0,0

9.6.3.2   Script File cbfreq

```
 1: % Example:  cbfreq
 2: % ~~~~~~~~~~~~~~~~
 3: % This program computes approximate natural
 4: % frequencies of a uniform depth cantilever
 5: % beam using finite difference and finite
 6: % element methods. Error results are presented
 7: % which demonstrate that the finite element
 8: % method is much more accurate than the finite
 9: % difference method when the same matrix orders
10: % are used in computation of the eigenvalues.
11: %
12: % User m functions required:
13: %    cbfrqnwm, cbfrqfdm, cbfrqfem, frud,
14: %    examplmo, animate, plotsave, read,
15: %    removvfil, genprint
16: %--------------------------------------------------
17:
18: clear; clf
19: fprintf('\n\n');
20: fprintf('CANTILEVER BEAM FREQUENCIES BY ')
21: fprintf('FINITE DIFFERENCE AND')
22: fprintf(...
23: '\n            FINITE ELEMENT APPROXIMATION\n')
24:
25: fprintf('\nGive the number of frequencies ')
26: fprintf('to be computed')
27: fprintf('\n(use an even number greater ')
28: fprintf('than 2)'), n=input('? > ');
29: if rem(n,2) ~= 0, n=n+1; end
30:
31: % Exact frequencies from solution of
32: % the frequency equation
33: wex = cbfrqnwm(n,1e-12);
34:
35: % Frequencies for the finite
36: % difference solution
37: wfd = cbfrqfdm(n);
38:
39: % Frequencies, modal vectors, mass matrix,
40: % and stiffness matrix from the finite
41: % element solution.
42: nelts=n/2; [wfe,mv,mm,kk] = cbfrqfem(nelts);
```

```
43: pefdm=(wfd-wex)./(.01*wex);
44: pefem=(wfe-wex)./(.01*wex);
45:
46: nlines=17; nloop=round(n/nlines);
47: v=[(1:n)',wex,wfd,pefdm,wfe,pefem];
48: disp(' '); lo=1;
49: t1=[' freq.      exact.         fdif.' ...
50:    '         fd. pct.'];
51: t1=[t1,'    felt.        fe. pct.'];
52: t2=['number     freq.          freq.' ...
53:    '         error    '];
54: t2=[t2,'    freq.         error   '];
55: while lo < n
56:    disp(t1),disp(t2)
57:    hi=min(lo+nlines-1,n);
58:    for j=lo:hi
59:      s1=sprintf('\n %4.0f %13.5e %13.5e', ...
60:                 v(j,1),v(j,2),v(j,3));
61:      s2=sprintf(' %9.3f %13.5e %9.3f', ...
62:                 v(j,4),v(j,5),v(j,6));
63:      fprintf([s1,s2])
64:    end
65:    fprintf('\n\nPress RETURN to continue\n\n')
66:    pause;
67:    lo=lo+nlines;
68: end
69: plotsave(wex,wfd,pefdm,wfe,pefem)
70: nfe=length(wfe); nmidl=nfe/2;
71: if rem(nmidl,2)==0, nmidl=nmidl+1; end
72: x0=zeros(nfe,1); v0=x0; w=0;
73: f1=zeros(nfe,1); f2=f1; f1(nfe-1)=1;
74: f1(nmidl)=-5;
75: xsav=examplmo(mm,kk,f1,f2,x0,v0,wfe,mv);
```

9.6.3.3 Function cbfrqnwm

```
1: function z=cbfrqnwm(n,tol)
2: %
3: % z=cbfrqnwm(n,tol)
4: % ~~~~~~~~~~~~~~~~~
5: % Cantilever beam frequencies by Newton's
6: % method.  Zeros of
7: %        f(z) = cos(z) + 1/cosh(z)
```

```
 8: % are computed.
 9: %
10: % n   - Number of frequencies required
11: % tol - Error tolerance for terminating
12: %         the iteration
13: % z   - Dimensionless frequencies are the
14: %         squares of the roots of f(z)=0
15: %
16: % User m functions called:  none
17: %-------------------------------------------------
18:
19: if nargin ==1, tol=1.e-5; end
20:
21: % Base initial estimates on the asymptotic
22: % form of the frequency equation
23: zbegin=((1:n)-.5)'*pi; zbegin(1)=1.875; big=10;
24:
25: % Start Newton iteration
26: while big > tol
27:    t=exp(-zbegin); tt=t.*t;
28:    f=cos(zbegin)+2*t./(1+tt);
29:    fp=-sin(zbegin)-2*t.*(1-tt)./(1+tt).^2;
30:    delz=-f./fp;
31:    z=zbegin+delz; big=max(abs(delz)); zbegin=z;
32: end
33: z=z.*z;
```

9.6.3.4 Function cbfrqfdm

```
 1: function [wfindif,mat]=cbfrqfdm(n)
 2: %
 3: % [wfindif,mat]=cbfrqfdm(n)
 4: % ~~~~~~~~~~~~~~~~~~~~~~~~~~
 5: % This function computes approximate cantilever
 6: % beam frequencies by the finite difference
 7: % method. The truncation error for the
 8: % differential equation and boundary
 9: % conditions are of order h^2.
10: %
11: % n       - Number of frequencies to be
12: %             computed
13: % wfindif - Approximate frequencies in
14: %             dimensionless form
```

```
15: % mat        - Matrix having eigenvalues which
16: %              are the square roots of the
17: %              frequencies
18: %
19: % User m functions called:  none
20: %---------------------------------------------
21:
22: % Form the primary part of the frequency matrix
23: mat=3*diag(ones(n,1))-4*diag(ones(n-1,1),1)+...
24:     diag(ones(n-2,1),2); mat=(mat+mat');
25:
26: % Impose left end boundary conditions
27: % y(0)=0 and y'(0)=0
28: mat(1,[1:3])=[7,-4,1]; mat(2,[1:4])=[-4,6,-4,1];
29:
30: % Impose right end boundary conditions
31: % y''(1)=0 and y'''(1)=0
32: mat(n-1,[n-3:n])=[1,-4,5,-2];
33: mat(n,[n-2:n])=[2,-4,2];
34:
35: % Compute approximate frequencies and
36: % sort these values
37: w=eig(mat); w=sort(w); h=1/n;
38: wfindif=sqrt(w)/(h*h);
```

9.6.3.5   Function cbfrqfem

```
1: function [wfem,modvecs,mm,kk]= ...
2:                     cbfrqfem(nelts,mas,len,ei)
3: %
4: % [wfem,modvecs,mm,kk]=
5: %                     cbfrqfem(nelts,mas,len,ei)
6: % ~~~~~~~~~~~~~~~~~~~~~~~~~~~~~~~~~~~~~~~~~~~~~~~~
7: % Determination of natural frequencies of a
8: % uniform depth cantilever beam by the Finite
9: % Element Method.
10: %
11: % nelts    - number of elements in the beam
12: % mas      - total beam mass
13: % len      - total beam length
14: % ei       - elastic modulus times moment
15: %            of inertia
16: % wfem     - dimensionless circular frequencies
```

```
17: %   modvecs - modal vector matrix
18: %   mm,kk   - reduced mass and stiffness
19: %             matrices
20: %
21: % User m functions called:  none
22: %-----------------------------------------------
23:
24: if nargin==1, mas=1; len=1; ei=1; end
25: n=nelts; le=len/n; me=mas/n;
26: c1=6/le^2; c2=3/le; c3=2*ei/le;
27:
28: % element mass matrix
29: masselt=me/420* ...
30:         [   156,    22*le,       54,   -13*le
31:            22*le,  4*le^2,  13*le,  -3*le^2
32:               54,   13*le,     156,  -22*le
33:          -13*le, -3*le^2, -22*le,   4*le^2];
34:
35: % element stiffness matrix
36: stifelt=c3*[ c1,   c2,   -c1,   c2
37:              c2,    2,   -c2,    1
38:             -c1,  -c2,    c1,  -c2
39:              c2,    1,   -c2,    2];
40:
41: ndof=2*(n+1); jj=0:3;
42: mm=zeros(ndof);   kk=zeros(ndof);
43:
44: % Assemble equations
45: for i=1:n
46:    j=2*i-1+jj; mm(j,j)=mm(j,j)+masselt;
47:    kk(j,j)=kk(j,j)+stifelt;
48: end
49:
50: % Remove degrees of freedom for zero
51: % deflection and zero slope at the left end.
52: mm=mm(3:ndof,3:ndof); kk=kk(3:ndof,3:ndof);
53:
54: % Compute frequencies
55: if nargout ==1
56:   wfem=sqrt(sort(real(eig(mm\kk))));
57: else
58:   [modvecs,wfem]=eig(mm\kk);
59:   [wfem,id]=sort(diag(wfem));
60:   wfem=sqrt(wfem); modvecs=modvecs(:,id);
61: end
```

9.6.3.6 Function frud

```
 1: function [t,x]= ...
 2:         frud(m,k,f1,f2,w,x0,v0,wn,modvc,h,tmax)
 3: %
 4: % [t,x]=frud(m,k,f1,f2,w,x0,v0,wn,modvc,h,tmax)
 5: % ~~~~~~~~~~~~~~~~~~~~~~~~~~~~~~~~~~~~~~~~~~~~~~~
 6: % This function employs modal superposition
 7: % to solve
 8: %
 9: %     m*x'' + k*x = f1*cos(w*t) + f2*sin(w*t)
10: %
11: % m,k     - mass and stiffness matrices
12: % f1,f2   - amplitude vectors for the forcing
13: %           function
14: % w       - forcing frequency not matching any
15: %           natural frequency component in wn
16: % wn      - vector of natural frequency values
17: % x0,v0   - initial displacement and velocity
18: %           vectors
19: % modvc   - matrix with modal vectors as its
20: %           columns
21: % h,tmax  - time step and maximum time for
22: %           evaluation of the solution
23: % t       - column of times at which the
24: %           solution is computed
25: % x       - solution matrix in which row j
26: %           is the solution vector at
27: %           time t(j)
28: %
29: % User m functions called:  none
30: %-----------------------------------------------
31:
32: t=0:h:tmax; nt=length(t); nx=length(x0);
33: wn=wn(:); wnt=wn*t;
34:
35: % Evaluate the particular solution.
36: x12=(k-(w*w)*m)\[f1,f2];
37: x1=x12(:,1); x2=x12(:,2);
38: xp=x1*cos(w*t)+x2*sin(w*t);
39:
40: % Evaluate the homogeneous solution.
41: cof=modvc\[x0-x1,v0-w*x2];
42: c1=cof(:,1)'; c2=(cof(:,2)./wn)';
```

```
43: xh=(modvc.*c1(ones(1,nx),:))*cos(wnt)+...
44:    (modvc.*c2(ones(1,nx),:))*sin(wnt);
45:
46: % Combine the particular and
47: % homogeneous solutions.
48: t=t(:); x=(xp+xh)';
```

9.6.3.7   Function examplmo

```
 1: function x=examplmo(mm,kk,f1,f2,x0,v0,wfe,mv)
 2: %
 3: % x=examplmo(mm,kk,f1,f2,x0,v0,wfe,mv)
 4: % ~~~~~~~~~~~~~~~~~~~~~~~~~~~~~~~~~~~~~~
 5: % Evaluate the response caused when a downward
 6: % load at the middle and an upward load at the
 7: % free end is applied.
 8: %
 9: % mm, kk - mass and stiffness matrices
10: % f1, f2 - forcing function magnitudes
11: % x0, v0 - initial position and velocity
12: % wfe    - forcing function frequency
13: % mv     - matrix of modal vectors
14: %
15: % User m functions called:  frud, animate, read
16: %-------------------------------------------------
17:
18: w=0; n=length(x0); t0=0;
19: s1=['\nEvaluate the time response ' ...
20:     'resulting from a '];
21: s2=['concentrated downward load at ' ...
22:     'the middle and '];
23: s3=['an upward end load.'];
24: while 1
25:    fprintf('\n'); fprintf(s1);
26:    fprintf('\n'); fprintf(s2);
27:    fprintf('\n'); fprintf(s3);
28:    fprintf('\n\n');
29:    fprintf('input the time step and ')
30:    fprintf('the maximum time ');
31:    fprintf('\n(0.04 and 5.0) are typical.');
32:    fprintf(' Use 0,0 to stop\n')
33:    [h,tmax]=read; disp(' ')
34:    if norm([h,tmax])==0, return; end
```

```
35:    [t,x]= ...
36:        frud(mm,kk,f1,f2,w,x0,v0,wfe,mv,h,tmax);
37:    x=x(:,1:2:n-1); x=[zeros(length(t),1),x];
38:    [nt,nc]=size(x); hdist=linspace(0,1,nc);
39:    plot(t,x(:,nc),'w');
40:    title('POSITION OF THE FREE END OF THE BEAM')
41:    xlabel('dimensionless time'),
42:    ylabel('end deflection'),
43:    disp('Press RETURN to continue'); pause;
44:    genprint('endpos');
45:    xc=linspace(0,1,nc); zmax=1.2*max(abs(x(:)));
46:    clf; mesh(xc,t,x); view(30,35);
47:    axis([0,1,0,tmax,-zmax,zmax])
48:    xlabel('x axis'); ylabel('time');
49:    zlabel('deflection')
50:    title(['BEAM DEFLECTION HISTORY ' ...
51:          'FOR VARYING TIME'])
52:    disp('Press RETURN to continue'); pause;
53:    genprint('cantsurf')
54:    titl='BEAM ANIMATION ';
55:    xlab='x axis'; ylab='displacement';
56:    animate(hdist,x,0.1,titl,xlab,ylab)
57: end
```

9.6.3.8 Function animate

```
 1: function animate(x,u,tpause,titl,xlabl,ylabl)
 2: %
 3: % animate(x,u,tpause,titl,xlabl,ylabl)
 4: % ~~~~~~~~~~~~~~~~~~~~~~~~~~~~~~~~~~~~~~~
 5: % This function draws an animated plot of data
 6: % values stored in array u.  The different
 7: % columns of u correspond to position values
 8: % in vector x.  The successive rows of u
 9: % correspond to different times. Parameter
10: % tpause controls the speed of animation.
11: %
12: % u       - matrix of values to animate plots
13: %             of u versus x
14: % x       - spatial positions for different
15: %             columns of u
16: % tpause  - clock seconds between output of
17: %             frames. The default is .1 secs
```

```
18: %              when tpause is left out. When
19: %              tpause=0, a new frame appears
20: %              when the user presses any key.
21: %  titl   - graph title
22: %  xlabl  - label for horizontal axis
23: %  ylabl  - label for vertical axis
24: %
25: % User m functions called:  none
26: %------------------------------------------------
27:
28: clf;
29: if nargin<6, ylabl=''; end;
30: if nargin<5, xlabl=''; end
31: if nargin<4, titl=''; end;
32: if nargin<3, tpause=.1; end
33:
34: [ntime,nxpts]=size(u);
35: umin=min(u(:)); umax=max(u(:));
36: udif=umax-umin;
37: uavg=.5*(umin+umax);
38: xmin=min(x); xmax=max(x);
39: xdif=xmax-xmin; xavg=.5*(xmin+xmax);
40: xwmin=xavg-.55*xdif; xwmax=xavg+.55*xdif;
41: uwmin=uavg-.55*udif; uwmax=uavg+.55*udif; clf
42: axis([xwmin,xwmax,uwmin,uwmax]); title(titl)
43: xlabel(xlabl); ylabel(ylabl); hold on
44:
45: for j=1:ntime
46:   ut=u(j,:); plot(x,ut,'w');
47:   if tpause==0
48:     pause;
49:   else
50:     pause(tpause);
51:   end
52:   if j==ntime, break, else, cla; end
53: end
54: genprint('cntltrac.ps')
55:
56: hold off; clf;
```

9.6.3.9 Function plotsave

```
1: function plotsave(wex,wfd,pefd,wfe,pefem)
```

```
 2: %
 3: % function plotsave(wex,wfd,pefd,wfe,pefem)
 4: % ~~~~~~~~~~~~~~~~~~~~~~~~~~~~~~~~~~~~~~~~~~~~
 5: % This function plots errors in frequencies
 6: % computed by two approximate methods.
 7: %
 8: % wex        - exact frequencies
 9: % wfd        - finite difference frequencies
10: % wfe        - finite element frequencies
11: % pefd,pefem - percent errors by both methods
12: %
13: % User m functions called:  genprint
14: %-------------------------------------------------
15:
16: % plot results comparing accuracy
17: % of both frequency methods
18: w=[wex(:);wfd(:);wfd];
19: wmin=min(w); wmax=max(w);
20: n=length(wex);wht=wmin+.001*(wmax-wmin);
21: j=1:n;
22: semilogy(j,wex,'-w',j,wfe,'--w',j,wfd,'-.w')
23: title('CANTILEVER BEAM FREQUENCIES');
24: xlabel('frequency number');
25: ylabel('frequency values');
26: xl=max(1.5,.1*n);
27: text(xl,wht,['solid <=> exact freq.    ' ...
28:              'dash <=> felt. freq.'])
29: text(xl,.3*wht,'.-.-. <=> fdif. freq.')
30: genprint('beamfreq');
31: disp('Press RETURN to continue'); pause;
32: plot(j,abs(pefd),'-w',j,abs(pefem),'--w')
33: title(['CANTILEVER BEAM FREQUENCY ' ...
34:        'ERROR PERCENTAGES'])
35: xlabel('frequency number');
36: eht=.8*max([pefd(:);pefem(:)]);
37: ylabel('percent frequency error');
38: xl=max(1.5,.1*n);
39: text(xl,eht,'solid <=> fdif. pct. error')
40: text(xl,.8*eht,'dash  <=> felt. pct. error')
41: disp('Press RETURN to continue'); pause;
42: genprint('frqerror');
```

Appendix A

References

[1] M. Abramowitz and I.A. Stegun. *Handbook of Mathematical Functions with Formulas, Graphs, and Mathematical Tables.* National Bureau of Standards, Applied Math. Series #55. Dover Publications, 1965.

[2] J. H. Ahlberg, E. N. Nilson, and J. L. Walsh. *The Theory of Splines and Their Applications.* Mathematics in Science and Engineering, Volume 38. Academic Press, 1967.

[3] J. Albrecht, L. Collatz, W. Velte, and W. Wunderlich, editors. *Numerical Treatment of Eigenvalue Problems*, volume 4. Birkhauser Verlag, 1987.

[4] E. Anderson, Z. Bai, C. Bischof, J. Demmel, J. Dongarra, J. Du Croz, A. Greenbaum, S. Hammarling, A. McKenney, S. Ostrouchov, and D. Sorensen. *LAPACK User's Guide.* SIAM, Philadelphia, 1992.

[5] Brian A. Barsky. *Computer Graphics and Geometric Modeling Using Beta-splines.* Computer Science Workbench. Springer-Verlag, 1988.

[6] Klaus J. Bathe. *Finite Element Procedures in Engineering Analysis.* Prentice-Hall, 1982.

[7] E. Becker, G. Carey, and J. Oden. *Finite Elements, An Introduction.* Prentice-Hall, 1981.

[8] Kellyn S. Betts. Math packages multiply. *CIME Mechanical Engineering*, pages 32–38, August 1990.

[9] K.E. Brenan, S.L. Campbell, and L.R. Petzold. *Numerical Solution of Initial-Value Problems in Differential-Algebraic Equations.* Elsevier Science Publishers, 1989.

[10] R. Brent. *Algorithms for Minimization Without Derivatives.* Prentice-Hall, 1973.

[11] P. Brown, G. Byrne, and A. Hindmarsh. VODE: A variable coefficient ODE solver. *SIAM J. Sci. Stat. Comp.*, 10:1038–1051, 1989.

[12] G. Carey and J. Oden. *Finite Elements, Computational Aspects*. Prentice-Hall, 1984.

[13] B. Carnahan, H.A. Luther, and James O. Wilkes. *Applied Numerical Methods*. John Wiley, 1964.

[14] Francois E. Cellier and C. Magnus Rimvall. Matrix environments for continuous system modeling and simulation. *Simulation*, 52(4):141–149, 1989.

[15] B. Char, K. Geddes, G. Gonnet, and S. Watt. *MAPLE User's Guide*, chapter First Leaves: A Tutorial Introduction to MAPLE. Watcom Publications Ltd., Waterloo, Ontario, 1985.

[16] R. V. Churchhill, J. W. Brown, and R. F. Verhey. *Complex Variables and Applications*. McGraw-Hill, 1974.

[17] Samuel D. Conte and Carl de Boor. *Elementary Numerical Analysis: An Algorithmic Approach*. McGraw-Hill, third edition, 1980.

[18] J. W. Cooley and J. W. Tukey. An algorithm for the machine calculation of complex fourier series. *Math. Comp.*, 19:297–301, 1965.

[19] Roy R. Craig Jr. *Structural Dynamics*. John Wiley & Sons, 1988.

[20] J. K. Cullum and R. A. Willoughby, editors. *Large Scale Eigenvalue Problems*, chapter High Performance Computers and Algorithms From Linear Algebra, pages 15–36. Elsevier Science Publishers, 1986. by J. J. Dongarra and D. C. Sorensen.

[21] J. K. Cullum and R. A. Willoughby, editors. *Large Scale Eigenvalue Problems*, chapter Eigenvalue Problems and Algorithms in Structural Engineering, pages 81–93. Elsevier Science Publishers, 1986. by R. G. Grimes, J. G. Lewis, and H. D. Simon.

[22] Philip J. Davis and Philip Rabinowitz. *Methods of Numerical Integration*. Computer Science and Applied Mathematics. Academic Press, Inc., second edition, 1984.

[23] Carl de Boor. *A Practical Guide to Splines*, volume 27 of *Applied Mathematical Sciences*. Springer-Verlag, 1978.

[24] J. Dennis and R. Schnabel. *Numerical Methods for Unconstrained Optimization and Nonlinear Equations*. Prentice-Hall, 1983.

[25] J. Dongarra, E. Anderson, Z. Bai, A. Greenbaum, A. McKenney, J. Du Croz, S. Hammerling, J. Demmel, C. Bischof, and D. Sorensen. LAPACK: A portable linear algebra library for high performance computers. In *Supercomputing 1990*. IEEE Computer Society Press, 1990.

[26] J. J. Dongarra, J. Du Croz, I. Duff, and S. Hammarling. A set of level 3 basic linear algebra subprograms. *ACM Transactions on Mathematical Software*, December 1989.

[27] J. J. Dongarra, J. Du Croz, S. Hammarling, and R. Hanson. An extended set of fortran basic linear algebra subprograms. *ACM Transactions on Mathematical Software*, 14(1):1–32, 1988.

[28] Jack Dongarra, Jeremy Du Croz, Iain Duff, and Sven Hammarling. A set of level 3 basic linear algebra subprograms. Technical report, Argonne National Laboratory, Argonne, Illinois, August 1988.

[29] Jack Dongarra, Peter Mayes, and Giuseppe Radicati di Brozolo. The IBM RISC System/6000 and linear algebra operations. Technical Report CS-90-122, University of Tennessee Computer Science Department, Knoxville, Tennessee, December 1990.

[30] J.J. Dongarra, J.R. Bunch, C.B. Moler, and G.W. Stewart. *LINPACK User's Guide*. SIAM, Philadelphia, 1979.

[31] A. C. Eberhardt and G. H. Williard. Calculating precise cross-sectional properties for complex geometries. *Computers in Mechanical Engineering*, Sept./Oct. 1987.

[32] W. Flugge. *Handbook of Engineering Mechanics.* McGraw-Hill, 1962.

[33] G. Forsythe and C. B. Moler. *Computer Solution of Linear Algebraic Systems.* Prentice-Hall, 1967.

[34] G. Forsythe and W. Wasow. *Finite Difference Methods for Partial Differential Equations.* Wiley, 1960.

[35] G. E. Forsythe, M. A. Malcolm, and C. B. Moler. *Computer Methods for Mathematical Computations.* Prentice-Hall, 1977.

[36] B. S. Garbow, J. M. Boyle, Jack Dongarra, and C. B. Moler. *Matrix Eigensystem Routines — EISPACK Guide Extension*, volume 51 of *Lecture Notes in Computer Science.* Springer-Verlag, 1977.

[37] C. W. Gear. *Numerical Initial Value Problems in Ordinary Differential Equations.* Prentice-Hall, 1971.

[38] James Gleick. *Chaos: Making a New Science.* Viking, 1987.

[39] G.H. Golub and J. M. Ortega. *Scientific Computing and Differential Equations: An Introduction to Numerical Methods.* Academic Press, Inc., 1992.

[40] G.H. Golub and C.F. Van Loan. *Matrix Computations.* Johns Hopkins University Press, second edition, 1989.

[41] D. Greenwood. *Principles of Dynamics.* Prentice-Hall, 1988.

[42] R. Grimes and H. Simon. New software for large dense symmetric generalized eigenvalue problems using secondary storage. *Journal of Computational Physics*, 77:270–276, July 1988.

[43] R. Grimes and H. Simon. Solution of large, dense symmetric generalized eigenvalue problems using secondary storage. *ACM Transactions on Mathematical Software*, 14(3):241–256, September 1988.

[44] P. Henrici. *Discrete Variable Methods in Ordinary Differential Equations*. Wiley, 1962.

[45] E. Horowitz and S. Sohni. *Fundamentals of Computer Algorithms*. Computer Science Press, 1978.

[46] Thomas J. Hughes. *The Finite Element Method — Linear Static and Dynamic Finite Element Analysis*. Prentice-Hall, 1987.

[47] Jagmohan L. Humar. *Dynamics of Structures*. Prentice-Hall, 1990.

[48] L.V. Kantorovich and V.I. Krylov. *Approximate Methods of Higher Analysis*. Interscience Publishers, 1958.

[49] W. Kerner. Large-scale complex eigenvalue problems. *Journal of Computational Physics*, 85(1):1–85, 1989.

[50] C. Lanczos. *Applied Analysis*. Prentice-Hall, 1956.

[51] Leon Lapidus and John Seinfeld. *Numerical Solution of Ordinary Differential Equations*. Academic Press, 1971.

[52] C. Lawson and R. Hanson. *Solving Least Squares Problems*. Prentice-Hall, 1974.

[53] C. Lawson, R. Hanson, D. Kincaid, and F. Krogh. Basic linear algebra subprograms for fortran usage. *ACM Transactions on Mathematical Software*, 5:308–325, 1979.

[54] Y. T. Lee and A. A. G. Requicha. Algorithms for computing the volume and other integral properties of solids, i. known methods and open issues. *Communications of the ACM*, 25(9), 1982.

[55] Itzhak Levit. A new numerical procedure for symmetric eigenvalue problems. *Computers & Structures*, 18(6):977–988, 1984.

[56] J. A. Liggett. Exact formulae for areas, volumes and moments of polygons and polyhedra. *Communications in Applied Numerical Methods*, 4, 1988.

[57] J. Marin. Computing columns, footings and gates through moments of area. *Computers & Structures*, 18(2), 1984.

[58] Leonard Meirovitch. *Analytical Methods in Vibrations*. Macmillan, 1967.

[59] Leonard Meirovitch. *Computational Methods in Structural Dynamics*. Sijthoff & Noordhoff, 1980.

[60] C. Moler and G. Stewart. An algorithm for generalized matrix eigenvalue problems. *SIAM Journal of Numerical Analysis*, 10(2):241–256, April 1973.

[61] C.B. Moler and C.F. Van Loan. Nineteen dubious ways to compute the exponential of a matrix. *SIAM Review*, 20:801–836, 1979.

[62] N.I. Muskhelishvili. *Some Basic Problems of the Mathematical Theory of Elasticity*. P. Noordhoff, Groninger, Holland, 4th edition, 1972.

[63] N.I. Muskhelishvili. *Singular Integral Equations*. P. Noordhoff, Groninger, Holland, 2nd edition, 1973.

[64] Z. Nehari. *Conformal Mapping*. McGraw-Hill, 1952.

[65] D. T. Nguyen and J. S. Arora. An algorithm for solution of large eigenvalue problems. *Computers & Structures*, 24(4):645–650, 1986.

[66] J. M. Ortega. *Matrix Theory: A Second Course*. Plenum Press, 1987.

[67] J. M. Ortega and W. C. Rheinboldt. *Iterative Solution of Nonlinear Equations in Several Variables*. Academic Press, 1970.

[68] B. Parlett. *The Symmetric Eigenvalue Problem*. Prentice-Hall, 1980.

[69] Mario Paz. *Structural Dynamics: Theory & Computation*. Van Nostrand Reinhold Company, 1985.

[70] R. Piessens, E. de Doncker-Kapenga, C.W. Uberhuber, and D.K. Kahaner. *QUADPACK: A Subroutine Package for Automatic Integration*, volume 1 of *Computational Mathematics*. Springer-Verlag, 1983.

[71] P. Prenter. *Splines and Variational Methods*. Wiley, 1975.

[72] William H. Press, Brian P. Flannery, Saul A. Teukolsky, and William T. Vetterling. *Numerical Recipes: The Art of Scientific Computing*. Cambridge University Press, 1986.

[73] J. R. Rice. *The Approximation of Functions, Volumes 1 and 2*. Addison-Wesley, 1964.

[74] R. J. Roark and W. C. Young. *Formulas for Stress and Strain*. McGraw-Hill, 1975.

[75] Scientific Computing Associates, Inc., New Haven, CT. *CLAM User's Guide*, 1989.

[76] N. S. Sehmi. *Large Order Structural Analysis Techniques*. John Wiley, New York, 1989.

[77] L. Shampine and M. Gordon. *Computer Solutions of Ordinary Differential Equations: The Initial Value Problem*. W. H. Freeman, 1976.

[78] B. T. Smith, J. M. Boyle, Jack Dongarra, B. S. Garbow, Y. Ikebe, V. C. Klema, and C. Moler. *Matrix Eigensystem Routines — EISPACK Guide*, volume 6 of *Lecture Notes in Computer Science*. Springer-Verlag, 1976.

[79] R. Stepleman, editor. *Scientific Computing*, chapter ODE-PACK, A Systemized Collection of ODE Solvers. North Holland, 1983. by A. Hindmarsh.

[80] G. W. Stewart. *Introduction to Matrix Computations*. Academic Press, 1973.

[81] G. Strang. *Introduction to Applied Mathematics*. Wellesley-Cambridge Press, 1986.

[82] G. Strang. *Linear Algebra and Its Applications*. Harcourt Brace Jovanovich, 1988.

[83] The MathWorks Inc. *MATLAB User's Guide*. The Math-Works, Inc., South Natick, MA, 1991.

[84] The MathWorks Inc. *The Spline Toolbox for Use With MATLAB*. The MathWorks, Inc., South Natick, MA, 1992.

[85] The MathWorks Inc. *The Student Edition of MATLAB For MSDOS Personal Computers*. The MATLAB Curriculum Series. Prentice-Hall, Englewood Cliffs, NJ, 1992.

[86] Charles Van Loan. A survey of matrix computations. Technical Report CTC90TR26, Cornell Theory Center, Ithaca, New York, October 1990.

[87] G. A. Watson, editor. *Lecture Notes in Mathematics*, volume 506, chapter An Overview of Software Development for Special Functions. Springer-Verlag, 1975. by W. J. Cody.

[88] D. W. White and J. F. Abel. Bibliography on finite elements and supercomputing. *Communications in Applied Numerical Methods*, 4:279–294, 1988.

[89] J. H. Wilkinson. *Rounding Errors in Algebraic Processes*. Prentice-Hall, 1963.

[90] J. H. Wilkinson. *The Algebraic Eigenvalue Problem*. Oxford University Press, 1965.

[91] J. H. Wilkinson and C. Reinsch. *Handbook for Automatic Computation, Volume II: Linear Algebra*. Springer-Verlag, 1971.

[92] H. B. Wilson. *A Method of Conformal Mapping and the Determination of Stresses in Solid-Propellant Rocket Grains*. PhD thesis, Dept. of Theoretical and Applied Mechanics, University of Illinois, Urbana, IL, February 1963.

[93] H. B. Wilson and K. Deb. Inertial properties of tapered cylinders and partial volumes of revolution. *Computer Aided Design*, 21(7), September 1989.

[94] H. B. Wilson and D. S. Farrior. Computation of geometrical and inertial properties for general areas and volumes of revolution. *Computer Aided Design*, 8(8), 1976.

[95] H. B. Wilson and S. Gupta. Beam frequencies from finite element and finite difference analysis compared using MAT-LAB. *Sound and Vibration*, 26(8), 1992.

[96] H. B. Wilson and J. L. Hill. Volume properties and surface load effects on three dimensional bodies. Technical Report BER Report No. 266-241, Department of Engineering Mechanics, University of Alabama, Tuscaloosa, Alabama, 1980. U.S. Army Engineer Waterways Experiment Station, Vicksburg, MS, 1980.

[97] Howard B. Wilson and Kalyanmoy Deb. Evaluation of high order single step integrators for structural response calculation. *Journal of Sound and Vibration*, 141(1):55–70, 1991.

[98] S. Wolfram. *A System for Doing Mathematics by Computer*. Addison-Wesley, 1988.

[99] C. R. Wylie. *Advanced Engineering Mathematics*. McGraw-Hill, 1966.

[100] D. Young and R. Gregory. *A Survey of Numerical Mathematics, Volume 1 and 2*. Chelsea Publishing Co., 1990.

Appendix B

List of MATLAB Routines with Descriptions

TABLE B.1. List of MATLAB Routines with Descriptions

Routine	Chapter	Description
adams2	4	Function to determine coefficients in the Adams type formulas used to solve $y'(x) = f(x, y)$.
adamsex	4	Driver program illustrating use of function **adams2** for determining coefficients in the explicit or implicit Adams type formulas used for differential equation solution.
allprop	5	Function to compute area, centroidal coordinates, and second moments of area (inertial moments) for a spline curve.
animate	9	Function to draw an animated plot of data values stored in an array.
arcprop	5	Function which computes area properties of a circular arc.
areaprop	5	Function used to determine geometrical properties of any area bounded by straight lines and circular arcs.
arearun	5	Driver program to compute the area, centroidal coordinates, moments of inertia, and product of inertia for any area bounded by straight lines and circular arcs.
		continued on next page

Routine	Chapter	Description
beamresp	9	Function to evaluate the time dependent displacement and bending moment in a constant cross-section simply-supported beam which is initially at rest when a harmonically varying moment is suddenly applied at the right end.
besf	5	Companion function for script runcases.
bisect	1	Determines a root of a function by interval halving.
boxrot	2	Function which applies rotation transformations to animate the motion of a cube rotating about a vertical axis.
boxrun	2	Driver program to animate the motion of a rotating cube.
cablemk	7	Function to form the mass and stiffness matrices for a cable.
cablenl	8	Driver program for numerical integration of the matrix differential equations for the nonlinear dynamics of a cable fixed at both ends.
cablinea	7	Program which uses modal superposition to compute the dynamic response of a cable suspended at one end and free at the other.
canimate	7	Function to draw an animated plot of cable deflection data values.
cbfreq	9	Driver program for computing approximate natural frequencies of a uniform depth cantilever beam using finite difference and finite element methods.
		continued on next page

Routine	Chapter	Description
cbfrqfdm	9	Function to compute approximate cantilever beam frequencies by the finite difference method.
cbfrqfem	9	Function to determine natural frequencies of a uniform depth cantilever beam by the finite element method.
cbfrqnwm	9	Function to determine cantilever beam frequencies by Newton's method.
cbpwf6	5	Function to compute base points and weight factors in a composite quadrature formula which integrates an arbitrary function by dividing the interval of integration into equal parts and integrating over each part using a Gauss formula requiring six function values.
chbpts	2	Function to compute points with Chebyshev spacing.
chopsine	6	Sample code extract.
cplot	3	Function to plot complex array data undistorted in a square window.
crclovsq	5	Script file defining a half annulus above a square which contains a square hole. Used as input for script file **arearun**.
crnrtest	5	Example showing effect of corner points on spline curve approximation of a general geometry.
deislner	7	Driver program using implicit second or fourth order integrators to compute the linearized dynamical response of a cable suspended at one end and is free at the other end.
		continued on next page

Routine	Chapter	Description
derivtrp	4	Function which computes coefficients to interpolate derivatives by finite differences.
dife	5	Function to differentiate evenly spaced data.
dvdcof	4	Function using divided differences to compute coefficients for polynomial interpolation by the Newton form of the interpolating polynomial.
dvdtrp	4	Function which performs polynomial interpolation using the Newton form of the interpolating polynomial.
elipprop	5	Driver program to compute the area, centroidal coordinates, and inertial moments of a rotated ellipse.
equamo	8	Function forming nonlinear equations of motion for a cable fixed at both ends and loaded only by gravity forces.
eventime	8	Function to compute cable position coordinates for a series of evenly spaced time values.
examplmo	9	Function which evaluates the response caused when a downward load at the middle and an upward load at the free end is applied.
expc	5	Companion function for script runcases.
fbot	9	Function used to solve Laplace's equation for a rectangle.
fftaprox	6	Sample code extract for Fourier series approximation.
		continued on next page

Routine	Chapter	Description
finidif	4	Program using truncated Taylor series to compute finite difference formulas approximating derivatives of arbitrary order which are interpolated at an arbitrary set of base points not necessarily evenly spaced.
flopex	3	Driver program used to determine the number of floating point operations required to perform several familiar matrix calculations.
floptest	3	Function to determine the FLOP counts for various matrix operations.
fnc	1	Function employed in root finding example.
forcresp	3	Sample code extract illustrating the solution of matrix differential equations.
fouaprox	6	Function for Fourier series approximation.
fouseris	6	Driver program to illustrate the convergence rate of Fourier series approximations derived by applying the FFT to a general function which may be specified either by piecewise linear interpolation in a data table or by analytical definition in a function given by the user.
fousum	6	Function to sum the Fourier series of a real valued function.
frud	9	Function employing modal superposition to solve $mx'' + kx = f_1 \cos(\omega t) + f_2 \sin(\omega t)$

continued on next page

Routine	Chapter	Description
genprint	2-3, 5-9	SYSTEM DEPENDENT function which saves a plot to a file.
gquad	5	Function used to evaluate the integral of a function over specified integration limits. The numerical integration is performed using a composite Gauss integration rule.
gquad10	5	Function used to evaluate the integral of a function over specified integration limits. The numerical integration is performed using a composite 10-point Gauss integration rule.
gquad6	5	Function used to evaluate the integral of a function over arbitrary integration limits. The numerical integration is performed using a composite 6-point Gauss integration rule.
grule	5	Function to compute Gauss base points and weight factors.
gtop	9	Function used in solution of Laplace's equation for a rectangle.
heat	9	Function to evaluate transient heat conduction in a slab which has the left end insulated and has the right end subjected to a temperature variation.
hmpf	5	Companion function for script runcases.
hsmck	6	Function to solve $m\, y_h''(t) + c\, y_h'(t) + k\, y_h(t) = 0$ subject to initial conditions of $y_h(0) = y_0$ and $y_h'(0) = v_0$.
imptp	6	Function defining a piecewise linear function resembling the ground motion of the earthquake which occurred in 1940 in the Imperial Valley of California.

continued on next page

Routine	Chapter	Description
jnft	6	Function computing integer order Bessel functions of the first kind computed by use of the Fast Fourier Transform (FFT).
laplarec	9	Program using Fourier series to solve Laplace's equation in a rectangle.
laprec	9	Function which sums the series solving Laplace's equation in a rectangle.
lineprop	5	Function to compute area property contributions associated with a polyline.
lintrp	4, 6, 8, 9	Function which performs piecewise linear interpolation through data. This function is an alternative to **table1**.
logf	5	Companion function for script runcases.
makcrcsq	5	Function used to create data for a geometry involving half of an annulus placed above a square containing a square hole.
makratsq	3	Driver program to create a rational function map of a unit disk onto a square.
matlbdat	4	Example illustrating the use of splines to draw the word MATLAB.
mbvp	3	Function used to solve a mixed boundary value problem for a function which is harmonic inside the unit disk, symmetric about the x axis, and has boundary conditions involving function values on one part of the boundary and zero gradient elsewhere.
		continued on next page

Routine	Chapter	Description
mbvprun	3	Driver program to analyze a mixed boundary value problem for a function harmonic inside a circle.
mckde2i	7	Function using a second order implicit integrator to solve the matrix differential equation $mx''+cx'+kx = f(t)$ where m, c, and k are constant matrices and $f(t)$ is an externally defined function.
mckde4i	7	Function using a fourth order implicit integrator to solve the matrix differential equation $mx''+cx'+kx = f(t)$ where m, c, and k are constant matrices and $f(t)$ is an externally defined function.
membrane	3	Function to compute the transverse deflection of a uniformly tensioned membrane subjected to uniform pressure.
mom	8	Function to compute the driving moment needed to produce an exact solution in forced pendulum response example.
motion	2	Function to animate the motion of a vibrating string.
ndbemrsp	9	Function to evaluate the nondimensional displacement and moment in a constant cross-section simply-supported beam.
oneovx	1	Function defining the integrand passed to **simpson** in root calculation example.
		continued on next page

Routine	Chapter	Description
output	4	Function to print results for **finidif**.
pinvert	8	Function defining the equation of motion for the pendulum example.
plft	9	Function used to solve Laplace's equation for a rectangle.
ploteasy	2	Plot function with a simple argument list.
plotjrun	6	Driver program to compute integer order Bessel functions.
plotmotn	8	Function to plot the cable time history.
plotsave	9	Function to plot errors in frequencies computed by two approximate methods.
pltundst	5	Function which creates an undistorted plot of a curve.
pltxmidl	8	Function to plot horizontal position of midpoint in nonlinear cable example.
polyplot	2	Program illustrating how the location of interpolation points affects the accuracy and smoothness of polynomial approximations.
polyterp	4	Function to interpolate through n data points using a polynomial of degree $n - 1$.
prun	8	Driver program for dynamics of an inverted pendulum analyzed using **ode45**.
qrht	9	Function used to solve Laplace's equation for a rectangle.
ratcof	3	Function determining coefficients to approximate a rational function.
		continued on next page

Routine	Chapter	Description
raterp	3	Function performing rational function interpolation using coefficients from function **ratcof**.
read	1, 2, 5, 6, 9	Function to interactively read up to 20 variables from single line.
readv	4	Function to input a vector of specified length.
recstrs	9	Function employing point matching to obtain an approximate solution for torsional stresses in a Saint Venant beam of rectangular cross section.
rector	9	Driver program for point matching solution of torsional stresses in a Saint Venant beam of rectangular cross section.
removfil	2-3, 5-9	SYSTEM DEPENDENT function used to delete a file if it exists.
rkdestab	8	Program to plot the boundary of the region of the complex plane governing the maximum step size which may be used for stability of a Runge-Kutta integrator of arbitrary order.
rnbemimp	9	Program analyzing an impact dynamics problem for an elastic Euler beam of constant cross section which is simply supported at each end.
rootest	1	Instructional program illustrating nested function calls. The base of natural logarithms (e) is approximated by finding a value of x which makes the chosen function equal zero.

continued on next page

Routine	Chapter	Description
runimpv	6	Driver program for the earthquake dynamics example.
runcases	5	Driver program for comparing results from several numerical integration methods.
setup	9	Function used in example involving the Laplace equation for a rectangle.
shftprop	5	Function to compute area properties for a set of axes rotated relative to the original axes.
shkbftss	6	Function used in earthquake example to determine the steady state solution of a scalar differential equation.
shkstrng	2	Function computing the motion of a string having one end fixed and the other end shaken harmonically.
simpson	1, 5	Function to integrate by Simpson's rule.
simpsum	5	Special Simpson's rule function concerning a matrix.
sine	6	Function specifying all or part of a sine wave.
sinetrp	4	Driver program illustrating cubic spline approximation of $\sin(x)$, its first two derivatives, and its integral.
sinfft	9	Function determining coefficients in the Fourier sine series of a general real valued function.
slabheat	9	Program illustrating the temperature variation in a one-dimensional slab with the left end insulated and the right end given a temperature variation $\sin(\omega t)$.

continued on next page

Routine	Chapter	Description
spc	4, 5	Function to compute a matrix containing coefficients needed to perform a piecewise cubic interpolation among data values. The output from this function is used by function **spltrp** to evaluate the cubic spline function, its first two derivatives, or the function integral from $x(1)$ to an arbitrary upper limit.
spcurv2d	4, 5	Function to tabulate points on a spline curve connecting data points on a general curve.
splaprop	5	Function to compute geometrical properties of an area bounded by a spline curve.
spltrp	4, 5	Function for cubic spline interpolation using a data array obtained by first calling function **spc**. This function also integrates and differentiates.
sqmp	2, 3	Function to evaluate the Schwarz-Christoffel transformation mapping $abs(z) \leq 1$ inside a square of side length two.
sqtf	5	Companion function for script runcases.
squarrun	2	Driver program to plot the mapping of a circular disk onto the interior of a square by the Schwarz-Christoffel transformation.
squarmap	2	Function to evaluate the conformal mapping produced by the Schwarz-Christoffel transformation mapping $abs(z) \leq 1$ inside a square.

continued on next page

Routine	Chapter	Description
stringmo	2	Driver program to illustrate motion of a string having one end subjected to harmonic oscillation.
sumser	9	Function used in a beam dynamics example to sum a double Fourier series.
trisub	3	Function to solve $LX = B$, where L is lower triangular.
udfrevib	7	Function to compute undamped natural frequencies, modal vectors, and time response by modal superposition.
ulbc	9	Function used in example on solution of Laplace's equation for a rectangle.
unsymerr	8	Function to compute an error measure in nonlinear cable dynamics example.

Appendix C

MATLAB Utility Functions

C.0.3.1   Function genprint

```
 1: function genprint(fname,append)
 2: %
 3: % genprint(fname,append)
 4: % ~~~~~~~~~~~~~~~~~~~~~~~
 5: % This function saves a plot to a file.
 6: %
 7: % fname  - name of file to save plot to
 8: % append - optional, if included plot is
 9: %            appended to file fname
10: %
11: % SYSTEM DEPENDENT ROUTINE
12: %
13: % User m functions called:  removfil
14: %-------------------------------------------------
15:
16: % Save to postscript file
17: if nargin == 1
18:    eval(['removfil ', fname, '.ps']);
19:    eval(['print -dps ',fname]);
20: else
21:    eval(['print -dps -append ',fname]);
22: end
```

C.0.3.2   Function lintrp

```
 1: function y=lintrp(x,xd,yd)
 2: %
 3: % y=lintrp(x,xd,yd)
 4: % ~~~~~~~~~~~~~~~~~
 5: % This function performs piecewise linear
 6: % interpolation through data defined by vectors
 7: % xd,yd. The components of xd are presumed to
 8: % be in nondecreasing order. Any point where
```

```
 9: % xd(i)==xd(i+1) generates a jump
10: % discontinuity. For points outside the data
11: % range, the interpolation is based on the
12: % lines through the outermost pairs of points
13: % at each end.
14: %
15: % x      - vector of values for which piecewise
16: %          linear interpolation is required
17: % xd,yd - vectors of data values through which
18: %          interpolation is performed
19: % y      - interpolated function values for
20: %          argument x
21: %
22: % User m functions required: none
23: %-----------------------------------------------
24:
25: x=x(:); xd=xd(:); yd=yd(:);
26: y=zeros(length(x),1);
27: xmin=min(x); xmax=max(x); nd=length(xd);
28:
29: if xmax > xd(nd)
30:    yd=[yd; yd(nd)+(yd(nd)-yd(nd-1))/ ...
31:       (xd(nd)-xd(nd-1))*2*(xmax-xd(nd)))];
32:    xd=[xd;2*xmax-xd(nd)]; nd=nd+1;
33: end
34:
35: if xmin < xd(1)
36:    yd=[yd(1)+(yd(2)-yd(1))/(xd(2)-xd(1))* ...
37:       (xmin-xd(1));yd];
38:    xd=[xmin;xd]; nd=nd+1;
39: end
40:
41: for i=1:nd-1
42:    xlft=xd(i); ylft=yd(i);
43:    xrht=xd(i+1); yrht=yd(i+1); dx=xrht-xlft;
44:    if dx~=0, s=(yrht-ylft)/dx;
45:      y=y+(x>=xlft).*(x<xrht).* ...
46:         (ylft+s*(x-xlft));
47:    end
48: end
49:
50: k=find(x==xd(nd));
51: if length(k)>0, y(k)=yd(nd); end
```

C.0.3.3 Function read

```
 1: function [a1,a2,a3,a4,a5,a6,a7,a8,a9,a10, ...
 2:           a11,a12,a13,a14,a15,a16,a17,a18, ...
 3:           a19,a20]=read(labl)
 4: %
 5: % [a1,a2,a3,a4,a5,a6,a7,a8,a9,a10,a11,a12, ...
 6: %  a13,a14,a15,a16,a17,a18,a19,a20]=read(labl)
 7: %~~~~~~~~~~~~~~~~~~~~~~~~~~~~~~~~~~~~~~~~~~~~~~~
 8: %
 9: % This function reads up to 20 variables on one
10: % line. The items should be separated by commas
11: % or blanks. Using more than 20 output
12: % variables will result in an error.
13: %
14: % labl               - Label preceding the
15: %                      data entry. It is set
16: %                      to '? ' if no value of
17: %                      labl is given.
18: % a1,a2,...,a_nargout - The output variables
19: %                      which are created
20: %                      (cannot exceed 20)
21: %
22: % A typical function call is:
23: % [A,B,C,D]=read('Enter values of A,B,C,D: ')
24: %
25: % User m functions required: none
26: %----------------------------------------------
27:
28: if nargin==0, labl='? '; end
29: n=nargout;
30: str=input(labl,'s'); str=['[',str,']'];
31: v=eval(str);
32: L=length(v);
33: if L>=n, v=v(1:n);
34:    else, v=[v,zeros(1,n-L)]; end
35: for j=1:nargout
36:   eval(['a',int2str(j),'=v(j);']);
37: end
```

C.0.3.4 Function readv

```
1: function [v,l]=readv(n)
2: %
3: % v=readv(n)
4: % ~~~~~~~~~~
5: % This function inputs a vector of length n.
6: % If fewer than n values are given, then the
7: % vector is padded with zeros.
8: %
9: %   n  - number of values to be input
10: %   v  - a vector of length n. If fewer than n
11: %        values were input, the final values
12: %        of v are set to zero
13: %   l  - the number of values actually read
14: %
15: % User m functions called:  none
16: %-----------------------------------------------
17:
18: str=input('? > ','s');
19: str=['[',str,']'];
20: v=eval(str); l=length(v);
21: if l>n, v=v(1:n); end
22: if l<n, v=[v,zeros(1,n-l)]; end
```

C.0.3.5 Function removfil

```
1: function removfil(fname)
2: %
3: % removfil(fname)
4: % ~~~~~~~~~~~~~~~
5: % This function removes file fname, if it
6: % exists
7: %
8: % SYSTEM DEPENDENT ROUTINE
9: %
10: % User m functions called:  none.
11: %----------------------------------------------
12:
13: if exist(fname)==2, eval(['!rm ',fname]); end
```

C.0.3.6 Function simpson

```
1: function ansr=simpson(funct,a,b,neven)
2: %
3: % ansr=simpson(funct,a,b,neven)
4: % ~~~~~~~~~~~~~~~~~~~~~~~~~~~~~~~
5: %
6: % This function integrates "funct" from
7: % "a" to "b" by Simpson's rule using
8: % "neven+1" function values.  Parameter
9: % "neven" should be an even integer.
10: %
11: % Example use:  ansr=simpson('sin',0,pi/2,4)
12: %
13: % funct    -  character string name of
14: %             function integrated
15: % a,b      -  integration limits
16: % neven    -  an even integer defining the
17: %             number of integration intervals
18: % ansr     -  Simpson rule estimate of the
19: %             integral
20: %
21: % User m functions called: argument funct
22: %-------------------------------------------
23:
24: ne=max(2,2*round(.1+neven/2)); d=(b-a)/ne;
25: x=a+d*(0:ne); y=feval(funct,x);
26: ansr=(d/3)*(y(1)+y(ne+1)+4*sum(y(2:2:ne))+...
27:     2*sum(y(3:2:ne-1)));
```

C.0.3.7 Function spc

```
1: function splmat=spc(x,y,i,v,icrnr)
2: %
3: % splmat=spc(x,y,i,v,icrnr)
4: % ~~~~~~~~~~~~~~~~~~~~~~~~~~
5: % This function computes matrix splmat
6: % containing coefficients needed to perform a
7: % piecewise cubic interpolation among data
8: % values contained in vectors x and y. The
9: % output from this function is used by function
10: % spltrp to evaluate the cubic spline function,
```

```
11: % its first two derivatives, or the function
12: % integral from x(1) to an arbitrary upper
13: % limit.
14: %
15: %   x    - vector of abscissa values arranged
16: %            in increasing order. The number of
17: %            data values is denoted by n.
18: %
19: %   y    - vector of ordinate values
20: %
21: %   i    - a two component vector [i1,i2].
22: %            Parameters i1 and i2 refer to left
23: %            end and right end conditions,
24: %            respectively. These equal 1, 2,
25: %            or 3 as explained below.
26: %
27: %   v    - a two component vector [v1,v2]
28: %            containing end values of y'(x)
29: %            or y''(x)
30: %
31: %            if i1=1: y'(x(1)) is set to v(1)
32: %                 =2: y''(x(1)) is set to v(1)
33: %                 =3: y''' is continuous at x(2)
34: %            if i2=1: y'(x(n)) is set to v(2)
35: %                 =2: y''(x(n)) is set to v(2)
36: %                 =3: y''' is continuous at x(n-1)
37: %
38: %            Note: When i1 or i2 equal 3 the
39: %                  corresponding values of vector
40: %                  v should be zero
41: %
42: % icrnr - A vector of indices identifying
43: %            interior points where slope
44: %            discontinuities are to be generated
45: %            by requiring y''(x) to equal zero.
46: %            Vector icrnr should have components
47: %            lying between 2 and n-1. The vector
48: %            can be omitted from the argument list
49: %            if no slope discontinuities occur.
50: %
51: % User m functions called:  none
52: %------------------------------------------------
53:
54: x=x(:); y=y(:);
55: n=length(x); a=zeros(n,1); b=a; c=a;
```

```
56: d=a; t=a; if nargin < 5, icrnr=[]; end
57: ncrnr=length(icrnr);
58: i1=i(1); i2=i(2);
59: v1=v(1); v2=v(2);
60:
61: % Form the tridiagonal system to solve for
62: % second derivative values at the data points
63:
64: n1=n-1; j=2:n1;
65: hj=x(j)-x(j-1); hj1=x(j+1)-x(j);
66: hjp=hj+hj1; a(j)=hj./hjp; wuns=ones(n-2,1);
67: b(j)=2*wuns; c(j)=wuns-a(j);
68: d(j)=6.*((y(j+1)-y(j))./hj1- ...
69:        (y(j)-y(j-1))./hj)./hjp;
70:
71: % Form  equations for the end conditions
72:
73: % slope specified at left end
74: if i1==1
75:    h2=x(2)-x(1); b(1)=2; c(1)=1;
76:    d(1)=6*((y(2)-y(1))/h2-v1)/h2;
77:
78: % second derivative specified at left end
79: elseif i1==2
80:    b(1)=1; c(1)=0; d(1)=v1;
81:
82: % not a knot condition
83: else
84:    b(1)=1;
85:    a(1)=hj(1)/hj(2);
86:    c(1)=-1-a(1);
87:    d(1)=0;
88: end
89:
90: % slope specified at right end
91: if i2==1
92:    hn=x(n)-x(n-1); a(n)=1; b(n)=2;
93:    d(n)=6.*(v2-(y(n)-y(n-1))/hn)/hn;
94:
95: % second derivative specified at right end
96: elseif i2==2
97:    a(n)=0; b(n)=1; d(n)=v2;
98:
99: % not a knot condition
100: else
```

```
101:    a(n)=1;
102:    b(n)=-(x(n-1)-x(n-2))/(x(n)-x(n-2));
103:    c(n)=-(x(n)-x(n-1))/(x(n)-x(n-2));
104: end
105:
106: % Adjust for slope discontinuity
107: % specified by lcrnr
108:
109: if ncrnr > 0
110:   zro=zeros(ncrnr,1);
111:   a(icrnr)=zro; c(icrnr)=zro;
112:   d(icrnr)=zro; b(icrnr)=ones(ncrnr,1);
113: end
114:
115: % Solve the tridiagonal system
116: % for t(1),...,t(n)
117:
118: bb=diag(a(2:n),-1)+diag(b)+diag(c(1:n-1),1);
119: if a(1)~=0, bb(1,3)=a(1); d(1)=0;    end
120: if c(n)~=0, bb(n,n-2)=c(n); d(n)=0; end
121: t=bb\d(:);
122:
123: % Save polynomial coefficients describing the
124: % cubics for each interval.
125:
126: j=1:n-1; k=2:n;
127: dx=x(k)-x(j); dy=y(k)-y(j); b=t(j)/2;
128: c=(t(k)-t(j))./(6*dx); a=dy./dx-(c.*dx+b).*dx;
129: int=(((c.*dx/4+b/3).*dx+a/2).*dx+y(j)).*dx;
130: int=[0;cumsum(int)]; n1=n-1;
131: ypn=dy(n1)/dx(n1)+dx(n1)*(2*t(n)+t(n1))/6;
132:
133: % The columns of splmat contain the
134: % following vectors
135: % [x,y,x_coef,x^2_coef,x^3_coef,integral_coef]
136:
137: splmat=[ ...
138:   [ x(1), y(1), a(1),   0,   0,   0     ];
139:   [ x(j), y(j), a,      b,   c,   int(j) ];
140:   [ x(n), y(n), ypn,    0,   0,   int(n) ] ];
```

C.0.3.8 Function spcurv2d

```
 1: function [xout,yout,sout]= ...
 2:           spcurv2d(xd,yd,nseg,ncrnr)
 3: %
 4: % [xout,yout]=spcurv2d(xd,yd,nseg,ncrnr)
 5: % ~~~~~~~~~~~~~~~~~~~~~~~~~~~~~~~~~~~~~~~~
 6: %
 7: % This function tabulates points xout, yout on
 8: % a spline curve connecting data points xd, yd
 9: % on the cubic spline curve
10: %
11: %   xd,yd      - input data points
12: %   nseg       - number of tabulation intervals
13: %                used per spline segment
14: %   ncrnr      - point indices where corners are
15: %                required
16: %   xout,yout  - output data points on the
17: %                spline curve. The number of
18: %                points returned equals
19: %
20: %                nout=(nd-1)*nseg+1.
21: %
22: % User m functions called:  spc, spltrp
23: %-----------------------------------------------
24:
25: nd=length(xd); sd=(1:nd)';
26: if nargin==2; nseg=10; end
27: if nargin<=3, ncrnr=[]; end;
28: nout=(nd-1)*nseg+1; sout=linspace(1,nd,nout);
29: if norm([xd(1)-xd(nd),yd(1)-yd(nd)]) < 100*eps
30:    yp=(yd(2)-yd(nd-1))/2;
31:    iend=[1;1]; vy=[yp;yp];
32:    xp=(xd(2)-xd(nd-1))/2; vx=[xp;xp];
33: else
34:    iend=[3;3]; vy=[0;0]; vx=vy;
35: end
36: matx=spc(sd,xd,iend,vy,ncrnr);
37: maty=spc(sd,yd,iend,vx,ncrnr);
38: xout=spltrp(sout,matx,0);
39: yout=spltrp(sout,maty,0);
```

C.0.3.9   Function spltrp

```
 1: function f=spltrp(x,mat,ideriv)
 2: %
 3: % f=spltrp(x,mat,ideriv)
 4: % ~~~~~~~~~~~~~~~~~~~~~~
 5: % This function performs cubic spline
 6: % interpolation using data array mat obtained
 7: % by first calling function spc.
 8: %
 9: % x      - the vector of interpolation values
10: %          at which the spline is to be
11: %          evaluated.
12: % mat    - the matrix output from function spc.
13: %          This array contains the data points
14: %          and the polynomial coefficients
15: %          needed to define the piecewise
16: %          cubic curve connecting the data
17: %          points.
18: % ideriv - a parameter specifying whether
19: %          function values, derivative values,
20: %          or integral values are computed.
21: %          Taking ideriv equal to 0, 1, 2, or
22: %          3, respectively, returns values of
23: %          y(x), y'(x), y''(x), or the integral
24: %          of y(x) from the first data point
25: %          defined in array mat to each point
26: %          in vector x specifying the chosen
27: %          interpolation points. For any points
28: %          outside the original data range,
29: %          the interpolation is performed by
30: %          extending the tangents at the first
31: %          and last data points.
32: %
33: % User m functions called:  none
34: %------------------------------------------------
35:
36: % identify interpolation intervals
37: % for each point
38: [nrow,ncol]=size(mat);
39: xx=x(:)'; xd=mat(2:nrow);
40: xd=xd(:); np=length(x); nd=length(xd);
41: ik=sum(xd(:,ones(1,np)) < xx(ones(nd,1),:));
42: ik=1+ik(:);
```

```
43: xk=mat(ik,1); yk=mat(ik,2); dx=x(:)-xk;
44: ak=mat(ik,3); bk=mat(ik,4);
45: ck=mat(ik,5); intk=mat(ik,6);
46:
47: % obtain function values at x
48: if ideriv==0
49:   f=((ck.*dx+bk).*dx+ak).*dx+yk; return
50:
51: % obtain first derivatives at x
52: elseif ideriv==1
53:   f=(3*ck.*dx+2*bk).*dx+ak; return
54:
55: % obtain second derivatives at x
56: elseif ideriv==2
57:   f=6*ck.*dx+2*bk; return
58:
59: % obtain integral from xd(1) to x
60: elseif ideriv==3
61:   f=intk+(((ck.*dx/4+bk/3).*dx+ ...
62:     ak/2).*dx+yk).*dx;
63: end
```

Index